THE *Oysters* OF *Locmariaquer*

BOOKS BY ELEANOR CLARK

FICTION

The Bitter Box
Baldur's Gate
Dr. Heart, A Novella, and Other Stories
Gloria Mundi

OTHER

Rome and a Villa
The Oysters of Locmariaquer
Eyes, Etc., A Memoir
Tamrart: 13 Days in the Sahara

FOR CHILDREN

The Song of Roland

TRANSLATION

Dark Wedding by Ramón Sender

THE OYSTERS OF LOCMARIAQUER

Eleanor Clark

With an Introduction by MARK KURLANSKY

AN ecco BOOK

HARPERPERENNIAL MODERNCLASSICS

NEW YORK • LONDON • TORONTO • SYDNEY

HARPERPERENNIAL MODERNCLASSICS

A hardcover edition of this book was published in 1964 by Pantheon Books, a division of Random House, Inc.

First Ecco edition published 1998.
First Harper Perennial Modern Classics edition published 2006.

Maps by Guy Fleming

Library of Congress Cataloging-in-Publication Data

Clark, Eleanor.
The oysters of Locmariaquer / Eleanor Clark ; introduction by Mark Kurlansky.– 1st Harper Perennial Modern Classics ed.
p. cm.
"An Ecco book."
Originally published: New York : Pantheon Books, 1964.
ISBN-10: 0-06-088742-7
ISBN-13: 978-0-06-088742-1
1. Oyster culture–France–Locmariaquer. 2. European oyster–France–Locmariaquer. 3. Locmariaquer (France)–Social life and customs. I. Title.

SH367.F7C53 2006
639.410944'13–dc22 2005044799

06 07 08 09 10 RRD 10 9 8 7 6 5 4 3 2 1

TO

Red

The oyster is a classical character.

—John R. Philpots

—Canst tell how an oyster makes his shell?
—No.
—Nor I neither.

—*King Lear*

On the zoological side, the books most exploited for this journey of interrogation have been *Oysters* by C. M. Yonge (London, 1960), and *Les Huîtres* by G. Ranson (Paris, 1951). The object of the drama related in Chapter Three is *Les Huîtres du Morbihan* by Pierre Dalido (Paris, 1948). To these works among many the author is greatly indebted. The American scientists of our time whose writings have been most illuminating on the animal in question are Paul Galtsoff, V. L. Loosanoff, and the late T. C. Nelson, to be read in various scientific journals and in the publications of the Fish and Wildlife Service, U. S. Department of the Interior.

Most exceptional and heartfelt thanks are due to Dr. Louis Marteil, of the Institut Scientifique et Technique des Pêches Maritimes in Auray. His generous help many times over the past three years has amounted to a one-sided "cultural exchange" dictated by nothing but kindness and courtesy. No government could have arranged it; no person could repay it. He is in no way to be associated with any errors there may be in the final product. To Professor Edward Deevey of the Department of Biology, Yale University, and to Peter Matthiessen the author is also deeply grateful, for corrections and suggestions and the charity of time spent on the manuscript. For any remaining faults they are of course not responsible either.

The nature of the book has made it unruly in at least one regard. Because of its own exigencies of tone, which may be obvious enough to serve as an apology, the rule of italics for scientific names has been broken.

INTRODUCTION

IT IS A CURIOUS AND, to writers, troubling fact that it is almost impossible to predict the work for which an author will be remembered. *The Oysters of Locmariaquer*, published in 1964 and winner of a 1965 National Book Award, is Eleanor Clark's most remembered book. This is somewhat surprising because Clark, a protégée of Katherine Anne Porter, was generally described as a fiction writer. In reality the oyster book was one of three such travel books. Her earlier book, *Rome and a Villa*, was also widely praised. She also wrote a memoir. Her fiction output was only slightly greater, a novella and short story collection and four novels.

Clark was married to Robert Penn Warren. It said so on every book flap and in every biography. Even today in online listings of her available books this is often the only information given about the author. Yet Robert Penn Warren, himself a National Book Award winner, was seldom identified as the husband of Eleanor Clark.

Her first novel, *The Bitter Box*, had been published in 1946. Nearly two decades later, when *The Oysters of Locmariaquer* won its stellar and immediate recognition, Clark was fifty-one years old and had not yet written the two novels that were to be her crowning achievements, *Baldur's Gate* (1970) and *Gloria Mundi* (1979). Both were set in small-town New England, which is where she spent most of her life.

But *The Oysters of Locmariaquer* is set in Brittany, the Celtic peninsula in the northwest of France. Clark had spent a summer in Brittany as a child and returned for several months to write this book. France in the 1960s was a very different country than France today. In Brittany there were few televisions, the prized possessions of the wealthy, people referred to as *"gros richards."* Oystermen did not own televisions. Today someone might pick up this book and brace themselves for that anglophone affliction of the mind–francophilia. But this book is not a precious description of the quaint and lovely peasants of rural France who do everything a bit lovelier–a book in the tradition of Alice B. Toklas. Clark embraced France without sentimentality, telling sad stories with grace, humor, and a sense of life's ironies.

The southern coast of Brittany has long been one of the poorest regions of France, one of the first to try to eek out a living on Parisian tourism, on the promotion of souvenirs reflecting colorful costumes and curious folkways praised by such Paris literati as Honoré de Balzac.

If the oystermen of Locmariaquer were a hard luck story, so were their oysters. The oyster is the *Ostrea edulis*, sometimes called the European flat oyster. These particular oysters are known as Belons after the mouth of the Belon river, which had long produced some of Europe's best oysters. By the time Clark was doing her research, the Belons of the Belon had already largely disappeared and nearby Locmariaquer was supplying the shipping point in Riec-sur-Belon.

Oysters, like wines, are to a large degree a reflection on where they are grown. North America's Atlantic Coast provides a wide range of oysters, including the fat, bland oysters of Louisiana and the Gulf of Mexico; Florida oysters; Chesapeake Bay oysters; Blue Points of Long Island; Wellfleets of Cape Cod; and the small, strong-flavored oysters from Nova Scotia to Labrador. But this assortment of shellfish that fill menus and delight gourmets–large, small, briny, sweet–are all biologically identical. They are the *Crassostrea virginica.* Variations in temperature, salinity, and available food account for the differences.

But there are considerable biological differences between *Crassostrea virginca* and *Ostrea edulis.* Principally there are tremendous differences between *Crassostrea* and *Ostrea,* which are different biological families, live in different conditions, and even reproduce differently.

They also look different. *Ostrea* is flat and round, whereas a *Crassostrea* is thicker and more oval shaped, even banana or broad-bean shaped when not cultivated.

Clark says, "The *Crassostrea* would seem to be the gluttons of the tribe." *Crassostrea* lives in brackish, less salty water of estuaries where it finds huge food supplies. The *Ostrea* needs saltier water and live further out. In comparing the climatic differences of the ranges, one fact seems obvious: *Crassostrea,* which can flourish from the subtropical waters of Florida and the Gulf of Mexico to the subarctic of Labrador, is a far more versatile, heartier animal than the *Ostrea,* which lives from southern Scandinavia to northern Spain. And that is why *Crassostrea*, not only the *Crassostrea virginica* of North America but the *Crassostrea angulata*, the Portuguese oyster, is taking over and the more delicate *Ostrea* is vanishing. As the *Ostrea* disappear, beds of *Crassostrea* are planted. Today on both coasts of America, in Europe, and in most of the world, *Crassostrea* is the dominant family. This is a sad fact not only because one more hallmark of European culture is vanishing but because the *Ostrea edulis* is an extraordinarily delicious oyster, an oyster as rough as a taste of the sea, yet so delicate a taste, as Montaigne said of Bordeaux oysters, "that it is like smelling violets to eat them."

More recently than Clark's book, Belons have been farmed in Maine, but a Maine *Ostrea edulis* no more resembles an *Ostrea edulis* from Brittany than a *Crassotrea virginica* from the Chesapeake resembles one from Nova Scotia. The oyster is part biology and part geography.

And so with both the oyster and the oystermen disappearing, there is a sorrowful fatalism to Eleanor Clark's book. "Perhaps," she suggests, "after all, like the backward culture that still makes it possible, like the vanishing forests or any other general beauty of America, it is just an anachronism." Both quaint and prescient the book remains relevant, though it is amusing to read about the threat of the ubiquitous Howard Johnson's taking over the world. The sadness would be unbearable except that the author's wit keeps readers chuckling in spite of themselves.

Eleanor Clark was once asked about the variety of her work and she said, "But it is all one voice, mine–and I don't disown it." Wise not to for it was the voice that seduced readers. Beyond the hardness of the life, the sadness of the story, there is a voice that is so engaging, so entertaining that it is clear this travelogue could only have been written by a writer who had long labored in her craft. Perhaps the most remembered line is: "Obviously, if you don't love life, you can't enjoy an oyster." But there are many others. On describing history after the first-century Roman naturalist Pliny, whom she called "more an egghead than a nature lover," she noted, "Ostreaologists, like most other ologists, tend to hurry over the next fifteen hundred years or so, the way you hurry down a lonesome road at nightfall." But no one hurries down this road in mid-twentieth-century Brittany. This is a book to be savored.

–Mark Kurlansky

THE *Oysters* OF *Locmariaquer*

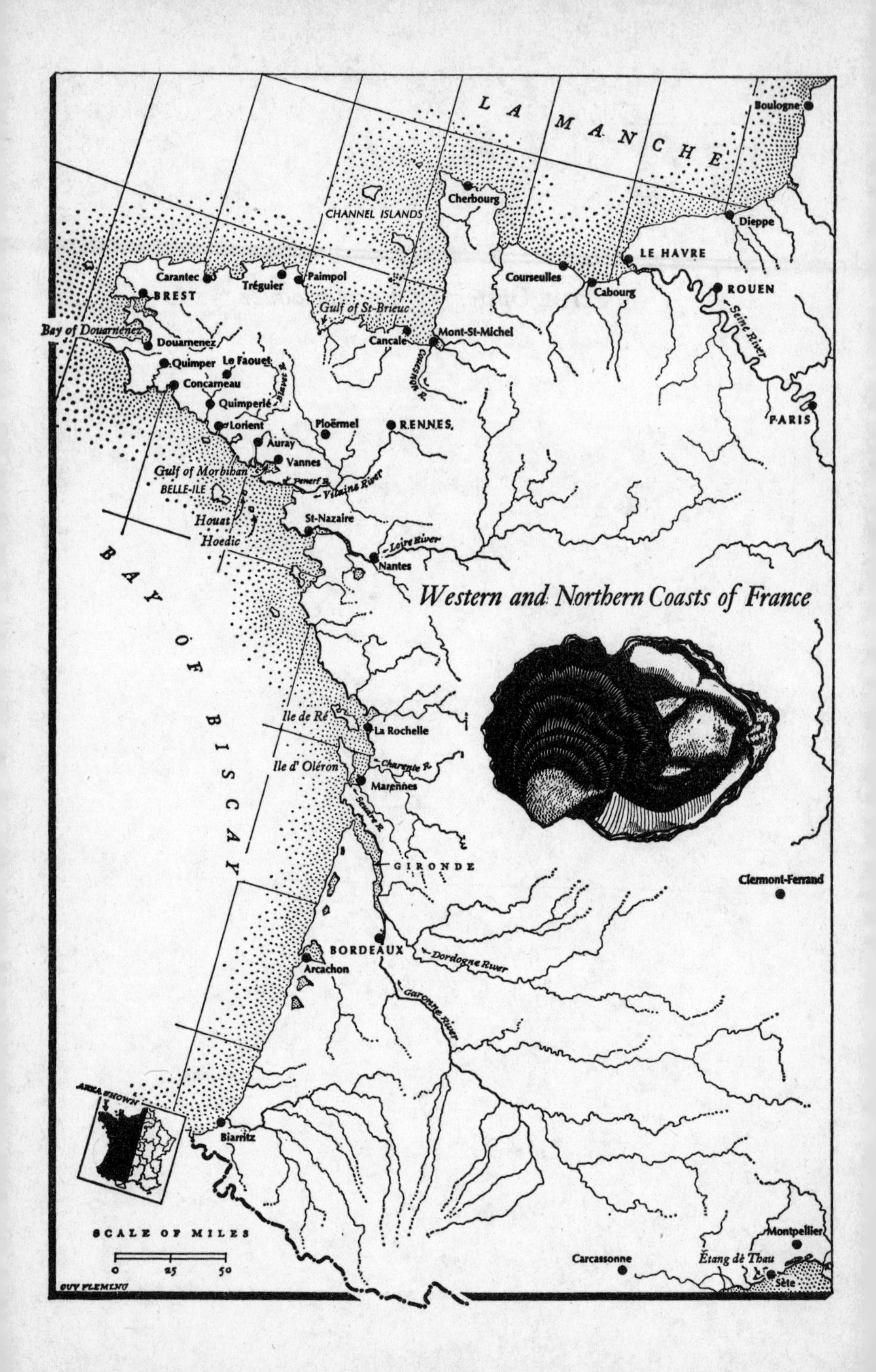
Western and Northern Coasts of France
LA MANCHE
Boulogne
Cherbourg
CHANNEL ISLANDS
Dieppe
LE HAVRE
Courseulles
Cabourg
ROUEN
Seine River
PARIS
Carantec
Tréguier
Paimpol
BREST
Gulf of St-Brieuc
Bay of Douarnenez
Douarnenez
Cancale
Mont-St-Michel
Couesnon R.
Quimper
Le Faouet
Concarneau
Blavet R.
Quimperlé
Lorient
Ploërmel
RENNES
Auray
Vannes
Gulf of Morbihan
Penerf R.
Vilaine River
BELLE-ILE
Houat
Hoedic
St-Nazaire
Loire River
Nantes
BAY OF BISCAY
Ile de Ré
La Rochelle
Ile d' Oléron
Charente R.
Marennes
Seudre R.
GIRONDE
Clermont-Ferrand
BORDEAUX
Arcachon
Dordogne River
Garonne River
Biarritz
AREA SHOWN
SCALE OF MILES
0
25
50
GUY FLEMING
Carcassonne
Étang de Thau
Montpellier
Sète

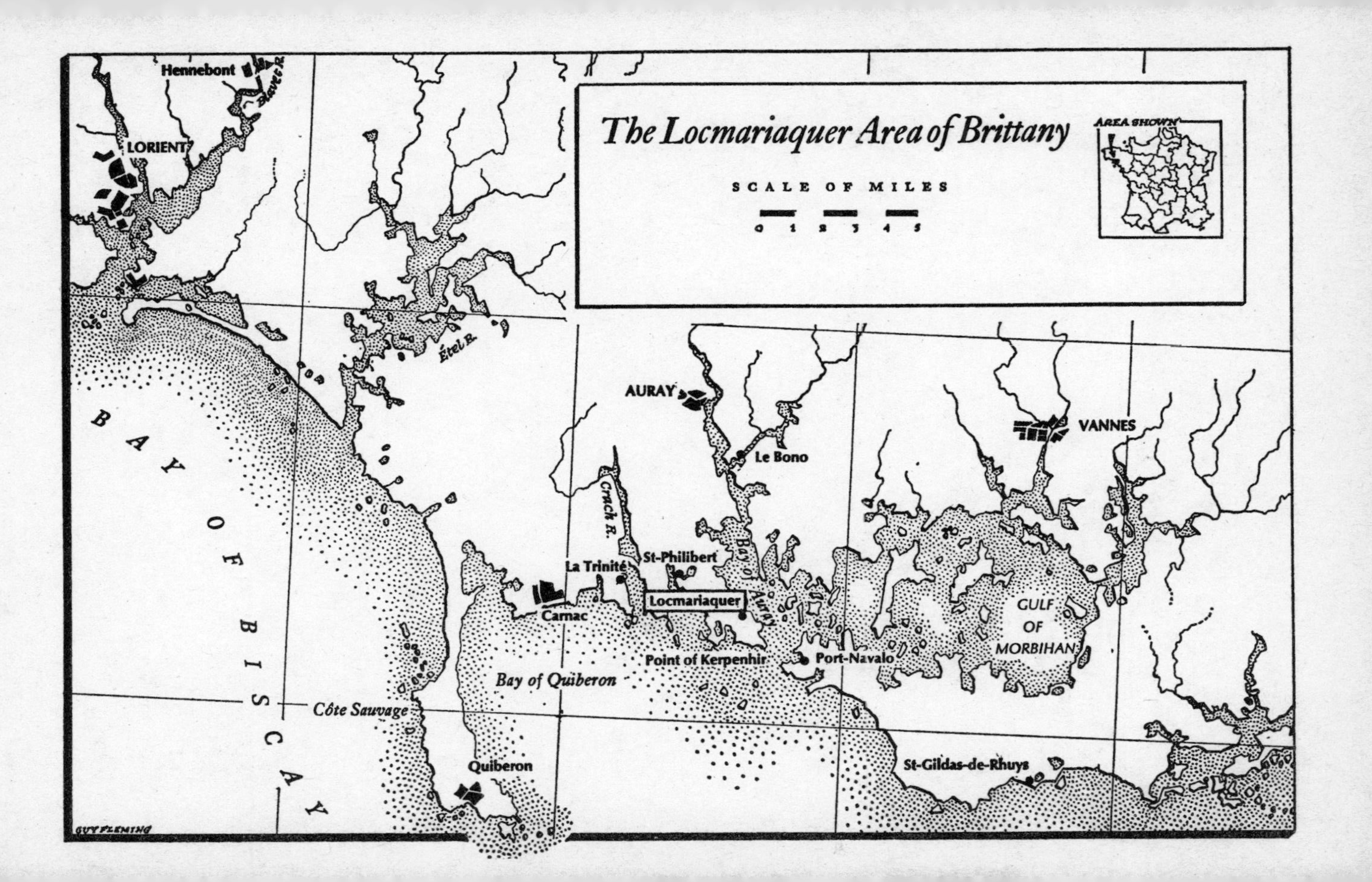
The Locmariaquer Area of Brittany
SCALE OF MILES
0 1 2 3 4 5
AREA SHOWN
Hennebont
Blavet R.
LORIENT
Étel R.
BAY OF BISCAY
AURAY
Le Bono
VANNES
Crach R.
La Trinité
St-Philibert
Bay of Auray
Locmariaquer
Carnac
GULF OF MORBIHAN
Point of Kerpenhir
Port-Navalo
Bay of Quiberon
Côte Sauvage
Quiberon
St-Gildas-de-Rhuys
GUY FLEMING

ONE

WHAT YOU NOTICE in the month of May is the tiles, like roof tiles but white, stacked by thousands at one point after another along the shore. There are twenty to thirty million of them altogether, all on what amounts to practically nothing on an ordinary road map of France, that is, a few miles of shoreline around the mouth of the Gulf of the Morbihan, from which this southern department of Brittany gets its name. It is Breton for Little (*bihan*) Sea (*mor*) and in fact the gulf has certain characteristics of a sea, with its own human and marine distinctions and its own violent little Gibraltar where the tides work like angry dragons and the German gun emplacements still stand. The oysters are bred in the gulf proper, along both banks of the estuary that branches from it, the River of Auray, and around the point in the other estuary at La Trinité. A brush-stroke and that is the end of it.

Not of oystering in general, of course. That is big business at many other points on the Brittany coast, and elsewhere. The oyster is not a creature of the open sea. It needs bays, coves, estuaries, and the coast of Brittany is the perfect geological crochet-work, after some terrible kitten had got through with it; if pulled out straight it would probably reach across the Atlantic and if you threw in the islands there is no telling where you would end up. So there are quite a few places where you can see the beautiful patterns of the *parcs* at low tide and the other shapes and paraphernalia of the trade.

They may have to do with either or both of the two European species of edible oyster. One is the plump Crassostrea angulata, called the Portuguese, which is cheaper and faster-growing and is not a leading character in this story, although in volume of production it is way ahead. It is a relative newcomer to France, having made its first appearance on the French coast, not in Brittany but farther south, in the middle of the last century. It can be marvelous enough by ordinary gastronomic standards, but the oyster of oysters, the most expensive, is the indigenous Ostrea edulis—called Armoricaine, from the ancient name for Brittany (*ar-mor:* by the sea), because Brittany raises 80 per cent of the world production of the species. They often appear as "Belon" on the menu, but that refers to their maturing phase, not their point of origin; in the region they are never called anything but *les plates*, the flat ones.

Almost all of them come from this one small pocket of the Morbihan, in a radius of a few miles out from the village of Locmariaquer, at the mouth, or maw, of the Gulf. At various stages of development they are all sent away to be fattened in other waters, some as late as three or four years old; almost all, and the few that stay to the end are skinny and undernourished. It is not a place for an oyster to grow up. The specialty and you might say obsession of the area is raising the babies of the species, *le naissain*, in English parlance "seed," "set," or "spat," the first stage in the long, hard, delicate process of producing an oyster fit to eat. In this the section has no rival worth mentioning. It is the world center for that one aspect of ostreiculture—or conchyliculture: it sounds very nice in French; and the baby Armoricaine has no rival in it. The concentration of shore-space and human effort toward the one single and singular end has a quality of dreams. "Tout le monde fait ça ici." The Portuguese oyster, which has been gradually moving farther north, might or might not do well in these waters, and might or might not drive out the *plate*; it has finally been admitted at two points in the Morbihan but right here there is still a law against it.

Not quite tout le monde. The proportion of men to women in the oyster-work is about one to ten; most of the men are away at sea, in the navy or merchant marine. There are some small-scale farmers, the usual little village bourgeoisie of storekeepers and so on, and in summer almost everybody does something else. The men may switch to sardine fishing up at Quiberon, the women get jobs in the vacation trade or the sardine canneries, also at Quiberon. Only the owner or a lone employee will be keeping watch over the parks all the time, and getting equipment ready for fall. It is the great reality just the same, only waiting to take over again when the last of the "Parigots" (Parisians) and other tourists have gone. To think of life is to think of oysters, all year round, almost as if you could hear all those millions of them breathing when the tide is out.

The end product is exquisite, the work grueling, and so badly paid, for everybody but a few of the owners, you wonder what keeps people so cheerful.

It is all on an artisan scale; the *chantiers* are small. There are over two thousand of them in the Morbihan, some just family affairs, a majority employing no more than ten or twelve workers at the height of the season, whether raising only the naissain or older oysters as well. The few big-shot producers, known as "les gros richards" because they are the rich of the region, with a TV in the house and a domestic servant in carpet slippers, are a little scornful of the seed business, whether or not they have parks of their own elsewhere. They buy from the small producers as many babies as they raise themselves and take them on from there, a year or two here, then away, then back for processing if this is the main base. But these stages of the work too, only a little less than the initial one, are a matter of hand, eye, muscle and sheer physical endurance. Very little of it is mechanized, and much of it could not be, without too great a loss both of oysters and of quality in those that would survive.

An uncanny degree of judgment is involved too, something beyond just knowing the business. It goes in the family, father to son; you have to be raised in it, at least so they say in the region and they can't name a successful producer who didn't come to it that way. It is the one business, they say, that an outsider can't make good in. That might be partly because there is no more room; the concessions are all taken. But what they are talking about is the sense of a thousand subtleties, imponderables of the sea, which in conjunction with the endless labor and patience go to make up the deep commitment associated with the word "métier."

The outcome is a little luxury item, of rather large economic consequence but no great importance to the world's nourishment. It should be. The oyster is very high in nutrition value, at least as much so as milk, but that is scarcely relevant as things stand because not enough people can afford it. So the whole point is flavor, and sociologically speaking, how can you justify that? Is it worth all the pain and trouble? Should it even be allowed?

You can't define it. Music or the color of the sea are easier to describe than the taste of one of these Armoricaines, which has been lifted, turned, rebedded, taught to close its mouth while traveling, culled, sorted, kept a while in a rest home or "basin" between each change of domicile, raked, protected from its enemies and shifting sands etc. for four or five years before it gets into your mouth. It has no relation at all to the taste, if there is one, of the usual U. S. restaurant oyster, not to mention the canned or frozen one. (No Armoricaines are canned, or frozen; there is no such business.) Or rather yes, it has the relation of love to tedium, delight to the death of the soul, the best to the tolerable if tolerable, in anything. Or say of French bread, the kind anybody eats in France, to . . . well, never mind. It is briny first of all, and not in the sense of brine in a barrel, for the preservation of something; there is a shock of freshness to it. Intimations of the ages of man, some piercing intuition of the sea and all its weeds and breezes shiver you a split second from that little stimulus on the palate. You are eating the sea, that's it, only the sensation of a gulp of sea water has been wafted out of it by some sorcery, and are on the verge of remembering you don't know what, mermaids or the sudden smell of kelp on the ebb tide or a poem you read once, something connected with the flavor of life itself . . .

You can eat them in the fancy restaurants of Nantes or Paris, or right out of the yard if you are lucky, or at almost any village café in the coast region. There, you can eat them at any time of year; in late June and July, the reproductive season, some are "milky" and horrid-tasting, but those are spotted when they are opened. The sign says BAR—CRÊPERIE—DÉGUSTATION D'HUÎTRES, and the word dégustation means what it says: not "consumption of" but "tasting," "savoring." It does not mean having a snack, with no suggestion beyond feeding your face. You are in the country of the art of good food, and this dégustation is very like what you do in an art gallery, unless your soul is lost; it is essential to be hungry but impermissible to be merely that; you have to take your time, the imagination must work; the first rule is to pay attention to what you are doing.

There is a certain expression that comes on a middle-to-upper income bracket Frenchman's face when he is about to déguster something really good, cheese, wine, any sort of culinary specialty, that starts out as a sudden interior break in the train of conversation. Silence; he is about to have a gastronomic experience. Then as the fork or glass nears his mouth, his eyes and ears seem to have blanked out; all is concentrated in the power of taste. There follows a stage when the critical faculties are gathering, the head is bent, eyes wander, lips and tongue are working over the evidence. At last comes the climactic moment of judgment, upon which may hang the mood of the meal and with it who knows what devious changes in the course of love, commerce or the body politic. The thing was poor or indifferent; the man shrugs, applies his napkin as though wiping out the whole experience, and goes on with what was interrupted, not quite relaxed; some sense of letdown, a slight disgruntlement lurks in the conversation. It was good, excellent, perfect, and oh what an expansion of frame and spirit; the chair will hardly hold him; he is not smiling, not just yet, but life is as he sits back gravely nodding, eager to look his companion and all the world in the eyes, and this time the napkin touches his lips like a chaste kiss, or a cleaning rag on an objet d'art.

It might be argued that this approach not only has a class angle to be considered but is old-fashioned. The Snack and the Snack Bar have appeared; the characters in the more fashionable French novels of recent years don't take any such view of food and in fact mostly don't seem to go in for it at all. Obviously, if you don't love life you can't enjoy an oyster. But ennui and social injustice are nothing to what the art of food has survived many times before in this country. The odors issuing from any boulangerie, from almost any provincial doorway at noon any day of the week can stir longings in an American, of which a sudden acuity of appetite is only the beginning.

Or consider the formidable directions printed on the menu of a café at La Trinité, the next town, on how to déguster ice cream: first warm each bite slightly in the spoon as you would warm a brandy glass, to bring out the flavor; eat nothing with it but one of two specified kinds of wafer or gauffrette, and so on.

In this fashion you approach, and ultimately consume, an oyster, and what comes upon you, beyond the shock of the flavor of the sea and as delicious, is there being no billboards along the roads. Only the ads painted on the ends of buildings at the edge of a town and they are quite gay; none of the other kind. That is not true on some big roads elsewhere in France, but it is throughout Brittany. As for architecture, the French can build as horridly as anybody else when they set their minds to it, and it may be true, as educated Bretons will tell you, that in another ten years the region will have gone the way of the world in that respect, as its little summer resorts mostly have, and the sections destroyed in the last war. Only the old towns and buildings, especially farm buildings and the obscure little chapels with their extraordinary sculpture, are beautiful. But they are still in such preponderance, and the lovely countryside from all the inland heights down to the various, sometimes demonic chinoiseries of the shore, is still so undefaced by our standards, there is such pleasure to the eye and the mind everywhere, it is as though a knot in your stomach, cramping the nerves and brain, were letting go. The sudden freedom from strain is bewildering. You are so used to ugliness—pervasive, scandalous, fly-by-night, brain-smashing ugliness, it takes a while to realize what has stopped hitting you. You can feel, you can think, you can enjoy food.

You wonder if we would be able to taste such an oyster at home even if we could have it, which we can't. Perhaps after all, like the backward culture that still makes it possible, like the vanishing forests or any other general beauty of America, it is just an anachronism. If Howard Johnson is the future then it certainly is, and will become extinct, as it nearly did once before within modern history, from a somewhat simpler cause.

However, the sea at least will stay, and some sort of oyster or oysters will probably survive in it here and there, even if lost to man.

Now about that happy feeling and the excitement of the faculties, it is not only aesthetic and cultural in source, although that is important enough. It has to do with the properties of the oyster, well-known to the ancient Romans, in its effects if not its chemistry. Aside from the oyster's nutritional and therapeutic values, and claims made for it by Casanova and many other enthusiasts through the centuries as an aphrodisiac—"a spur to the spirit and to love," as it may well be—it is apparently the best of all known stimulants to appetite, as well as digestion. This is attributed by scientists to the very high sodium chloride content of its flesh and especially of its liquids, which sets the gastric juices to flowing at a great rate, if a person is in a condition to have any. "Oyster dear to the gourmet," wrote Seneca, "beneficent Oyster, exciting rather than sating, all stomachs digest you, all stomachs bless you!"

THERE is a face and manner and way of being in Locmariaquer, which remains the queen of the oyster tiles in spite of some stiff competition from several points nearby, that is not quite like anything else even twenty miles away. There is a different world, prim and gigantic, at the haras in Hennebont, state stud-farm for the great race of Breton work-horses; and another, lusty and piratical-looking, around the big tuna fleet at Étel; there is a Quiberon swagger that goes with sardines and may also be subsidized by the mayor in midsummer as a tourist attraction; and a brooding air around the inland fishermen with their little flat-bottom boats and huge nets like Saint Peter's in the lonely brackish channels back of Auray, where each man works for himself.

The little world dedicated to the baby *plate*, while distinctly Breton, is distinct in Brittany. It is itself.

The term "Belon," which in the 1890's still meant oysters that had actually matured at the mouth of the Belon River in Finistère, was the subject of a lawsuit in Paris a few years ago. An effort was made to stop its use for oysters that had merely been sent to Riec-sur-Belon for shipment, to pick up the prestige of the name, or had not been there at all. It was discovered that only a tiny percentage of oysters called Belon could actually have come from there, so the court decided to sanction what it couldn't stop and let the term apply to all Armoricaines. Around Locmariaquer they don't like this too well, at least they don't let you forget that if it weren't for their naissain, Belon would have been out of the running long ago. "They all come from here."

You can call it local pride but you wouldn't be saying much. What matters is all the life and memory and unity of people in a place that can make for such pride, and that as subtly as from the taste of the oyster, voices of some ancient pleasure and reverence for a second disturb your compilation of facts.

Legends and a lot else went out with sailing ships. Still there is something you keep listening for. And why is it that nobody breaks up those wrecked hulks of boats with their scarey ribs?

ANOTHER mystery of the landscape is the four horses mirrored so beautifully for the moment in the brackish pond, below the old dyke. There are no riders for pleasure in these parts, yet these are nothing like work-horses. There is one of them, nearly two tons' worth, one of the great *traits* sired out of the haras at Hennebont, with the stumpy legs and chest like a mountainside, heaving across another field nearby. The farmer, as powerful in his kind, has a perfectly good tractor but still needs the horse for seeding in moist ground and some other work. His name is Giannot. His beautiful and still young-looking wife, who will be at it all day too and won't be looking young much longer, will appear on the next holiday serene as a fashion model in high heels and a suit that might be from Paris, but right now and most of the time is plodding along in sabots, bag of a dress and great white shawl, more like a model for Millet, behind the horse. Their children have just crossed the dyke on their way two miles to the nuns' school, only one old enough to have a bike, the others on foot; except for one old couple, everybody else around is in oyster-work and they have all gone off too, so the sounds of doves around the hedges and a seagull over the pond are tremendous. The gloomy pink mansion above the dyke, with closed shutters and big stone basins of the same period as the dyke, which appear as the tide ebbs there, used to be a farmhouse like any other. It was bought at one time by a man from Nantes who had the crazy idea of raising Portuguese oysters there, so he was driven out and the basins, half ruined now, were never used again.

In the field below, which in the days before the dyke used to be under the sea, the four wild long-legged horses, necks arched and slim, are frozen for a moment, listening, in their lovely pose beside the pond. It seems really too celestial for them to have grouped themselves just that way, against the water and the clouds, but then there is a lot that seems a little beyond nature in the early morning or evening in this light, with its strange pearly tricks and distances, even in a mist. The mood and the pose break suddenly; the horses in a mad impulse of gaiety scatter to the four corners, in four streams of light, as though the four reflections had sprung from the pond and were the reality.

Sometimes the menhirs and dolmens don't look quite solid either, although they are so many and so old and so heavy. They are the official mystery, and distinction, of the area; there are more than on any other patch of Europe and some of the biggest. Over three thousand menhirs in one field at Carnac, a few miles away. A menhir is a phallus-shaped rock on end, something to do with sun and fertility worship, nobody quite knows. A dolmen is a roofed tomb thing made of several such rocks and a few of the most famous are strewn around the pastures of Locmariaquer, awesome yet domesticated, and vastly boring, except when all of a sudden in mist or moonlight one of them is really there. Or when you have left the region and become aware of missing them; see one see all, pretty nearly, but it turns out they have an active part somehow in the play of land and sea and time that pulls the mind so, in this place.

There are just these two big things, prehistoric megaliths and baby oysters. It is not one of the fertile farm sections, the farming is poor and hard and on the wane; not many girls as good-looking as Giannot's wife would be willing to marry a farmer there any more.

The retreating tide pulls a scattering of lonely human figures after it off the dry land, as always in daytime. There is no oystering and no *vase*, the dark sea mud we have no proper word for, off this particular point, seaward from the pink mansion, and the few wild mussels to be found are on the rock shelves another finger around, more exposed. Here the bottom is gravelly, with occasional flat stretches of sharp-surfaced broken rock in a wallow of algae. There are several kinds of *fruits de mer* there but it is hard work getting them; the stooped silent figures, each with knife and basket, seem isolated by more than the distances between them; it is as though human language were cast off every day when the tide called. The tides are fairly modest, nothing like the great drama of the Normandy shore or the Bay of Saint-Brieuc on the north coast. Still they are long enough, a half-mile or more in the long tides, and the flux is part of the medium of life; nothing is to be understood, or seen, apart from it.

Most of the figures remain until the hunting grounds are covered again, and may appear basket on lap on the next market day in Auray, on one of the benches reserved for this category of food—the little clams, the *palourdes*, to be spotted by two tiny holes in the coarse sand, crayfish from the tidal ponds, diminutive sea snails. Some of the oyster women make a fairly good living this way in summer and prefer it to hiring themselves out as domestics to an inferior race, from Paris.

Françoise, old and crippled, who lives with her mother in the little stone hut nearest the shore, can't stay to the end and would have no way of getting to market even if she could fill her basket. Her slow, limping return brings wild yaps of excitement from her flea-ridden toy-sized dog, and a gelatinous quiver to the eye cavities of the mother, ninety-six years old, who has been waiting or rather deposited under a tree for the morning, like a huge damp mattress forced into a garment of black cotton. Françoise with her misshapen hip and her walking stick is not quick enough for the palourdes. The catch as usual is a small assortment of coarser shell creatures, which are quickly cut from their shells and fed into the soft avid opening that has appeared in the face of the mattress. But the old lady is still somebody after all, visions of young unvanquishable handsomeness haunt the rugged cut of the face, to a closer view. She turns stubborn, closes the protoplasmal aperture, pretends she is not hungry and that in every disintegrating fiber of her ancient bulk she is not craving the little gobbets of tonic from the sea. "Eat, eat," she says in Breton, the croak retaining nevertheless its old command, and with a brusque nod, for she is not quite blind, indicates her daughter's mouth, also the little bitch who has to nurse a litter of puppies the size of escargots and has been following the course of the gobbets, in despair. The eight or ten chickens, the family's whole remaining fortune, peck hectically at the shells. The nuns buy a few eggs there once in a while, for charity, or the priest in Auray for a special reason, or some passer-by who knows the story.

Long ago the old lady's uncles owned two farms, including the pink mansion. But so many tragedies have struck since then, in such a vengeful singling out of victims by the forces of misfortune, the last only ten years ago, anything that far back wouldn't count. Sometimes the mother still speaks of one of her sons killed in the last war, and will weep quietly; she has the power to make that distinction, among all her losses. But mostly, although Françoise's leaner, more nervous face might have been sculpted by the village knife-grinder, her sweetness of smile and capacity for interest in all sorts of things having no effect on the savage set of its suffering, mostly the two of them seem to have moved beyond any possibility of anguish, as though absorbed still living into the phases of the moon. Small things worry or please them. The noon church-bells clanging across the fields are pleasant, not as if it were three in the afternoon and meant a funeral. The schools and the oysteryards are letting out for lunch. For a little while the landscape will be astir again with voices and bicycles.

The people come past the fallen giant dolmen, the Table des Marchands, where the gypsies camp, and back over the dyke. Not as many as in the morning. Most of the children this far from the center have lunch at school, where at least in the two clerical schools they go in for aimless chasing and scrabbling in the gravel yard, like so many molecules under a microscope, for lack of any play equipment. The state school has a play gym with swings, and some of the luckier children own rubber balls; a doll or toy truck, even the cheapest, is a rarity. Yvette's two children are the only ones crossing the dyke at this hour. She collects them on her way from the chantier, one of the two biggest, over the other side of *le bourg*, as the center is called. A village is any little group of dwellings on the outskirts, so Locmariaquer includes a great many, each with its name and the names mostly have a *Ker* in them, the Breton for "house" or "place." The nuns are of the order of Kermaria and probably this village beyond the dyke, with its little chapel to Saint Peter, would have been Kerpierre but for the sound of it, instead of Saint-Pierre. Actually there are two villages there, a long stone's throw apart. Françoise's hut and the pink mansion and other ex-farm come under a different name.

Yvette lives between the two, in a tiny new white house beside the field where the four horses are. Except in summer, her family and the Giannots are the whole population of Saint-Pierre proper, and there is a little stiffness of rival beauties between her and Giannot's wife. They have the same broad cast of feature but Yvette is taller and there is more of a fling to her glance and her smile, which is a smile in depth, right out of her real self although as quick as if it were meaningless; she makes a largesse of life everywhere she passes on her bike, so of course is married to a handsome no-good, one of the worst drunks around. When sober he works for the other big oysteryard, of which her brother is foreman. She is twenty-eight and has been "in the oysters" since she was fourteen. Once when her husband was beating her he split her head open and their little girl Alice was so frightened she won't sleep there any more and has to live in the big farmhouse with her grandparents, but she goes home with her mother at lunchtime and helps her water the flowers and the horses, which are not theirs, and do whatever else there is time for, until they have to go back to town.

There is something peculiar about these horses at noon. All their fantastic element and even their horsiness seems to have left them. They are sprawled full length on the ground like dogs, wide awake, not even in the shade. Perhaps they are unhinged by their idyllic life, without any work to do; or somewhere in their memories there may be a smell of fear, which makes them skittish and debilitates them.

THE women have left the most grotesque part of their work costumes at the chantier—the huge rubber boots, gloves, rainsacks or sunbonnets of newspaper if there is sun, and may have changed back to skirts. However most of the spring work, up to the laying of the tiles, doesn't require pants. There is an uncommon number of beauties among them and an uncommon lack of grimness in all, although they work for 1.50 francs or about 35 cents an hour. The dark gypsy women, unkempt and serpentine at the door of their trailer truck, with their trinkets and exotic herbs to sell, heaven knows to whom around here at this time of year, regard them with mild, habitual scorn, in an ancientness of opposition that makes them sisters if they only knew it.

The energy of these women is prodigious. There are children, cattle, a vegetable garden and everything else to see to before and after eight hours of oystering, work as hard and in winter as painful as any in the world. In the long northern twilight they will be out with horse, child, possibly an aged grandparent, planting an acre of potatoes or cabbages, or be doing the week's wash at the well, in a stone tub that somebody carved out of a solid piece of granite a couple of hundred years ago. A family couldn't live on such earnings; there have to be chickens and rabbits, a pig. Yet there is hardly a woman who doesn't find time to grow flowers.

Along with stones, cows and the mutations of shoreline changing your sense of distance every hour, the countryside is most distinguished by flowers. Scribbled in flowers, made of flowers. In the spring there is such an explosion of wild ones everywhere, including the farm walls and roofs, and such a harmony of the delicate and the gross through all the changes of the composition, you would happily settle for those. A bang of yellow all over the place (juniper) gets replaced circus-fashion by Red, White and Blue (poppies, cornflowers, Queen Anne's lace) and that by a scatter-rug arrangement of mauve (tiny wild carnations). The senses reel; it's true, they do, they get quite mixed up; color and sound and smell, until you approach a barnyard, are all one pleasure, so no wonder you are hungry afterwards. This will confirm the thesis that an attractive landscape, even more than oysters, is the best of all appetizers, and the lack of it a national hazard through indigestion. The music is a throaty burbling of doves in all the ancient impenetrable hedges, the kind that spring up around Sleeping Beauties and which are one of the marks of all Brittany, and more stunningly from skylarks spiraling to heaven by their bootstraps of song, until they stop and in wild silence shoot themselves back to earth. The smell of honeysuckle is often way over your head too, since the paths are evidently the same ones used by the original Celts and have gone on being worn deeper all the time, while their thorny walls grew higher.

But there are all the other flowers too, that people work over, around every little new house no matter how poor, or old, old stone one where the inhabitants and their cows live under one roof and the footing outside is of mingled manure and ground mollusc-shells, and in every possible space in the bourg—especially roses, which seem to grow like Jack's beanstalk under these fleecy indeterminate skies. There must be some magic for them in the precise degree of salt in the air, or the gentle hide-and-seek weather which is rarely one thing or another very hard or very long. There are prize specimens at every turn, along with a huge gamut of other garden flowers, as though everybody had plenty of leisure time and were doing it for a hobby.

The national campaign for flowers, *La France Fleurie*, seems to apply as little in the region as the other on the same scale, against alcoholism. In the one province where no grapes grow, flowers and drunks are indigenous; no mere campaign could encourage one or discourage the other.

There is involved in all this, along with the specific trademarks of Locmariaquer and its environs, some general mystery of character shared with the rest of Brittany, pride of place, pride of being and of being in a certain rapport with the sea, that a foreigner can't fathom and the rest of the French can't either. It fascinates them. They are always writing about it or around it, often in denigration or else the opposite, not calmly anyway: the mists, the myths, the megaliths, the druids and dark woods of history, the sailors and their sea, the famous *pardons* or religious festivals in which it would appear that Bretons have retained a certain innocence of relation to God that the rest of France and the world mourns for itself, and the coiffes, costumes and other picturesqueness that have also slipped out of the world and are slipping in Brittany too. Every tourist leaves a little more of herself than she (these being largely feminine matters) picks up, everywhere she sets foot. Nowadays a woman under forty may wear a coiffe for a pardon, or if she is a waitress, in Finistère, but wouldn't be caught dead in one at any time in the Morbihan. The manufacturers of coiffe materials, and even of sabots although they are still much in use, are gradually closing down, and there are few old men left who know how to construct a thatched roof; luckily the slate used instead, as the old roofs collapse, is a handsome material too. Among moneyed city people the movement has even started to thatch their beach villas, at huge cost; at least there is one such roof over in Carnac, which would look a lot more genuine if it were fake.

Fashion models are blooming now among the menhirs, in Brittany. Most of this has happened since World War II, together with a strenuous and sometimes embarrassing revival of bagpipes, folk dancing, folklore, etc., as if the word "folk" used that way weren't a death warrant in itself.

Even so, it is true that tradition, or local identity, is dying far harder here than most places. Among older village and country women, not so old either that they wouldn't be "girls" in the U. S., the coiffe is still more or less uniform, at least in some sections. You see a great many of them, determined rather mysteriously by one's *coin*, "corner"—in this case the *coin d'Auray*, which has a simple flat coiffe, not like the startling lace stovepipes fifty miles away, around Concarneau, but in essence no less startling, for it means that people absolutely belong in one place and no other and if the going is rough they will not think of trying their luck somewhere else but will dig in deeper, identify more, not less, with the place, like Françoise in her hut by the shore. The corner is stronger than the *I*, and its symbolic headgear, however gay and colorful in some districts, has at least one thing in common with the coiffes of nuns. These too stand for the total acceptance of a bond and a condition, you might say a prenatal pact, in pride, the very opposite of resignation.

Even the sailors return each to his corner, to marry, to loaf, to die. They will bring home a monkey from Madagascar but never a Chinese or American wife, and hardly ever a French one who is not Breton, and seldom a Breton from more than twenty miles away. The girls and young wives living in the double drudgery of oyster-work and the stone barnyards most of them go home to will make things dim for tourists twenty years from now, but can't discard what the coiffe means so easily.

Anywhere in America with their looks they would be off to the city being starlets, secretaries, airline hostesses, and not all of Brittany is immune that way either. A lot of port towns and dreary limping upland ones far from the sea, without any communal dignity of occupation, provide Paris with prostitutes, and the people of one town, Gourin in the exact middle of the peninsula, have an even more dramatic peculiarity. Almost every family for ten miles around has at least one member in, or returned from, America, with the strange cultural consequence that a system of loudspeakers pours jazz at deafening pitch through every street in Gourin all day long. Just the same, a Breton is always a Breton, going or staying, sober or drunk; the prodigals may well be more coiffed and bound in their way than the women of Locmariaquer, who keep Ostrea edulis moving to the oysteryards of England, Scotland, northern Brittany and elsewhere, by the not at all simple fact of their never having thought for a minute, or so it seems, of getting out.

And no doubt it is because the priests and nuns of Brittany are of the same people, the same families and place and sense of what is and is to be, that there seems to be an abnormal amount of saintly, if always practical, dedication among them. They work as hard as anyone else; you can't imagine most of them ever finding time to be sick or tired; they are neither contemplative nor distinguished by any French esprit; have not retired from the world but are more in the driving thick of it than anybody else, having in their hands, along with nursing, recreation for young and old and plenty of other work, the education of two-thirds of the children. In education beyond the age of fourteen, the church's share is probably much higher.

Hence in part the famous conservatism of the province, putting it mildly, although the phenomenon, as in any place with a strong streak of passion for autonomy in modern times, both feeds and is fed by some furious contradictions; a bygone autonomy being the same in that respect as a brand-new one, say in Africa. The favorite bones of "martyrs" in the Auray neighborhood are not of church martyrs although they are on show in a church, but of a group of royalists massacred in Auray in 1795 after their attempt at a comeback from England; a great Breton hero remaining nevertheless one who was decapitated for fighting the crown not so long before. In Paris they tend to speak of Bretons as pigheaded and not very bright. But what strength, what lack of meanness or confusion of spirit, what indomitable, reckless goodness there is in the faces of some of these nuns and village priests, to whose characters you feel revolution is not more alien than Rome. For they are poor too, their lives are very hard, and at least one or two in almost every little place really care what is happening to people around them, all the time.

This doesn't, around Locmariaquer, put their hearts on the side of the *gros richards*, and as among the embattled artichoke farmers of the Côtes-du-Nord, might make for some painful paradoxes if the oyster-workers should ever start asking for their share. So far they haven't, and since the owners also work hard, and only a few are all that rich, and a great many do the actual physical labor of the business alongside their employees, there is to date no more than a delicate balance between grumbling over the pay and pride in the product, or perhaps more, in the process.

"Tout de même, c'est un beau métier."

Among other things, they might mean that they are like the fisherman of another country, who was secretly a merman at night and would have died if he had been kept away from the sea.

TWO

A THRILL of suspense runs through the region in the middle of June. Everyone is waiting for the word; even the storekeepers have a new keenness of eye for the weather and toward the port where the boats will pass. All the millions of tiles are ready. They have to be or the year is lost, so for the last week or so there has been nothing to do but load the first barge at each yard, and wait.

The white tiles you have been seeing for several miles around are "collectors." They are the surface on which the oyster larvae will attach themselves, and they have to be put out, in the right places naturally, and also at exactly the right time. If they are placed too early they become coated with *vase* and the oyster specks, which are invisible at that stage, can't hang on; if too late, the larvae are dispersed and killed. It was a guessing game in the old days, under threat of ruin, but finally the state stepped in. In a little side-street office and laboratory in Auray, a Docteur-ès-Sciences in marine biology and one assistant test samples of the various waters around, every day from early June. It is from there the word comes: the temperature and rate of oyster spawn are right; the tiles must be placed.

The date will differ by a week or ten days depending on the spot, but in the interior of the Gulf and right at its mouth, at Locmariaquer, there is little or no collecting, for a combination of reasons. From the hundreds of chantiers there the tiles have to be carried out through the strait to different points up the shore between Saint-Philibert and Quiberon, a bargeload a day for two weeks or more, until the job is done.

All of those start the same morning, at three, four, five, depending on the tide, each boat towing its ghostly load of ten or twelve thousand tiles at the end of an immense length of cable. People far inland are awakened by the roar of the motors, and those near the shores run out to watch. They will be used to it in a day or two, as in every other year of their lives, but that first morning the excitement is overpowering every time, as if the slow procession were really what it looks like in the first nacreous light, rounding the Point of Kerpenhir. Not a movement of armies, it only sounds like that and the German dugouts and pillboxes every few hundred yards along the dunes on the seaward side remind you of it. It is more as if the souls of all the dead and living, all forgiven and packaged alike in white, were about to be peacefully dumped somewhere up the coast.

It may be partly the time span of the oyster that gives such transcendence to the sight. Not, of course, the life span of the individual oyster; whatever that might be ideally, accidents and disease bring it to an average of about fifteen years. Nor are the hundred or so species now extant of particularly awesome age. With one or two possible but not proven exceptions they appeared on earth at the same time as man, at the beginning of the Pleistocene period, upwards of half a million years ago but probably not more than four times that by farthest present conjecture. But the oyster abounded on the planet long before that; it antedates man, and mountains and the shapes of continents as we know them, by a great many millions of years, and in some cases has changed since in ways not perceptible to the ordinary human eye.

But to be practical first, the beautiful movement of barges around Locmariaquer is part of a method that has never been in use in the United States, except for a few special experiments.

Whether American oyster-production, which has been steadily falling off over the last hundred years, can survive without some equivalent of it is another question.

It is a matter of breeding, as against merely fishing, or harvesting, from natural beds. The latter can be done either by dragging from boats in deep water, or on foot in beds that uncover at low tide, but since that kind of bed has been largely exterminated, offshore dragging is in the main the only system. In the United States, as in Holland and some other places, it has become a highly industrialized business; small-scale private oystering, which used to give a certain distinguishable human character to many points on our shores—Staten Island, for instance; even Bridgeport, Connecticut—has effectively vanished, under the pressure of competition when not from the demise of the oyster.

There are four stages in the cultivation of Ostrea edulis: (1) production, from the collecting of larvae, or *captage du naissain*, through babyhood, to an age officially called eighteen months but that is actually two years; (2) *l'élevage*, raising, until the oyster is three or four years old; (3) *l'affinage*, a further refinement of quality, over a period of a few months to a year; and (4) *l'expédition*, meaning not just shipping but a complicated set of procedures to get the oyster ready for market. Leaving out a few experiments and special situations, the whole cycle is carried out in areas that are exposed either twice a day or at least at the big tides. Some parts of the procedure are much the same as for the Portuguese, even at the breeding stage; collectors, of another kind, are used for that too at Arcachon and neighboring islands, farther south. But there are some important differences in the constitution and needs of the animal to begin with, and hence in techniques and human attitudes.

The Portuguese, like the American eastern oyster, is not of the genus Ostrea; it is a Crassostrea, formerly called Gryphaea. The so-called Pearl Oyster is not an oyster at all; neither are the creatures called Saddle Oysters and Thorny Oysters. However the pearl confusion has a basis in human yearnings, so is far more important.

To be an oyster you have to be of course a mollusc, and a bivalvular one, of the ostreid family, meaning acephalous, lamellibranchiate, monomyarian, asyphonic and inequivalvular;

which is to say: headless, with a crescent of tight little ruffles for breathing tubes or gills, with one muscle instead of two holding your valves together, lacking a water spout, and with one half of your shell quite different from the other. Like the other bivalves you are entirely enclosed in a sack called the mantle, a marvelous piece of material which not only fabricates every speck of shell you will ever have but also is in charge of all your sensory contact with the outside world. Unlike most other bivalves you have the peculiar habit of lying on your so-called left, that is lower or cupped, valve, heaven knows why or when the habit was acquired. You will usually find yourself described along with clams, scallops, etc. as a pelecypod, meaning with a hatchet-shaped foot, in distinction to the stomach-footed gastropods such as snails and limpets, or the head-footed branch of the phylum that includes the octopus, called cephalopods. This, as an oyster, you should not tolerate, since after the larval stage you have no more foot than head. However you have excellent nerves, stomach, liver, etc. and the best gills in the kingdom. By another caprice of terminology, instead of monomyarian you are likely to be called anisomyarian, meaning with two unequal adductor muscles; this too, like pelecypod, relates to a larval and ancestral condition, not to your adult anatomy.

There are some more delicate requirements we can pass over for the moment.

The hundred-odd living species of creatures satisfying all the above have been classified in various ways since science got on the job, but from recent studies of the larval shell are now usually grouped in three genera—Ostrea, Crassostrea, and Pycnodonta. This last, like most lamellibranchs or ruffle breathers but no other oyster, has its rectum running through its heart; this is considered rather unsophisticated. As between the first two genera, the shells are far more distinct than the bodies, as can easily be seen in the smoother and nearly round *plate* as versus the bumpy and usually elongated Portuguese. Not that the difference is so obvious with all species, also there can be big variations both individually and between oysters of the same species in different places, which has caused quite a confusion of names in some cases. On the functional side, only the Ostrea species are larviparous, i.e., lay their eggs inside their valves and keep them there for a week after fertilization.

If you cast your eggs on the waters to be fertilized, you are called not unmaternal but merely oviparous, and are by that fact either a Crassostrea or a Pycnodonta. Differentiations of species, not to get on to races within species, are a great deal harder and far from finished. "It seems that the problems will be solved only with that of the structure and functioning of living matter, of protoplasm, that is of life itself." (G. Ranson)

There's a solemn question; and another, to return to that, is the sweep of the thing in time, a wonderful journey if we can manage it. We rarely do manage it. It seems the time sense of the human race, all but the most backward, has been out of kilter for the last century or so. The old feel of it is gone and the new one hasn't anywhere near caught up with what we know since what we know got expanded too suddenly, like objects in dreams. We are not sick, just insecure, with a bad footing in between recorded human history, which looks either too small or too big for comfort depending on one's mood, and some vague apprehension of astronomy. The oyster will help to put us straight. Beneficent Oyster! all minds bless you, all minds can comprehend you, after a fashion.

There are some 500 known fossil species of oyster, dating from different geological eras, especially from the exuberant Mesozoic, when the earth rested for some millions of years from its earlier convulsions and animal life burst forth with a new flourish. More than half of these species are of the Cretaceous period at the end of the era, coinciding more or less with the high point, or beginning of decline, of the dinosaurs. This time, 70 to 100 million years ago, seems to have been a most propitious time for our friend, as it was for other molluscs, some of them gigantic, and fish, and land vertebrates, notably weird-shaped reptiles up to 180 feet long, with a little horsey head on top of a long neck. But the oyster is said to have appeared, that is to have differentiated from earlier or protolamellibranchs, another 100 million years or so before that, or some time after the span of the monstrous forests and the formation of coal, the samples from Carboniferous times not having been identified with certainty as true oysters. Most, if perhaps not all, of these ancestral types at least of the Mesozoic era, lived in fairly deep and salty water, a characteristic retained by only a few of their descendants outside the Pycnodonta, which in that way as in others are the most archaic of present-day oysters. Present-day meaning roughly within the last half-million years, or two or three ice ages back; there have been few changes since then.

The changes, such as they were, from earlier species occurred on certain occasions if that is the right word, coinciding with the great geological epochs, by a process affecting most if not all forms of life, and at this point and this one alone we have to admit that the oyster is not much more of a comfort than anything else. Just how a species gets itself genetically transformed, and by what geological or other developments, is an unsettling question, especially as the process is said to be "sudden," a relative term to be sure. But we will have to put up with it. At the end of the Cretaceous period, we are told, a creature called Pycnodonta flabelliformis disappears and the next thing we know we are dealing with P. martinsi instead, which in turn with the Oligocene gets rid of itself in suddenly giving birth to P. squarrosa, just as suddenly replaced in recent time by P. hyotis. The same with their cousins, the immediate parent of Ostrea edulis being the Miocene-Pliocene O. lamellosa, through an evolution similar in rather widely separated parts of the earth at the same time. (G. Ranson) In the case of one or two of the Ostrea, a question has been raised of their having possibly escaped the general renovation of fauna marking the beginning of man's time on earth; a fascinating possibility.

A few of these kinds of oysters had and have a wide distribution around the world. One such cosmopolite is the tropical P. hyotis, as far-flung in habitat as coral reefs and usually found in conjunction with them, in the Indian Ocean, the South Pacific, the Caribbean, the west coast of Africa, etc. But these are rare. Most oysters are more specific in their needs and so more strictly localized, living conditions being hardly ever exactly duplicated in coastal waters, without speaking of man's doings. The key factors are salt, or density of the water, temperature, and the nature of the bottom. The species differ a good deal in all three respects but there is none that can do without salt; a fresh-water creature that looks like an oyster and is called one in local parlance, as in certain equatorial rivers, will turn out to be of some other family.

In general the Pycnodonta are in the saltiest water, then the Ostrea, then the Crassostrea. Among the latter, the American eastern oyster, C. virginica, requires slightly fresher water than the Portuguese, C. angulata; neither can thrive in deep water far from the coast, to the extent of forming natural banks, except where strong fresh-water currents of one sort or another exist. They are more apt to be found in large quantities in the intertidal zone, or strip of shore uncovered regularly by the tides, and can even exist so far upstream as to be in absolutely fresh water twice a day, though by dint of closing tight at those times. They are also far more resistant to extremes of temperature. Winters of bitter cold, aberrations of tides causing too long exposure in summer heat, excessive rains, are all conditions far more likely to be mortal to the *plate* than to the Portuguese, or to our virginica. The Crassostrea can also stand encroachments of slime that would kill a colony of Ostrea, through a difference in eating habits and partly because of their shape, which allows them to stand on end with their behinds poking out of the mud, sometimes supported by tiers of shells of the dead in the same position. The more rounded *plate* has no such reach to save it from big alluvial deposits. It favors a harder bottom, rock, gravel, shell or whatever. Its natural banks will be at the mouths of smaller, less turbid rivers, or rivers that are more like fjords as they are in the neighborhood of Locmariaquer; and the banks, unlike the oystermen's parks, will be at the extreme limit of low water where they rarely uncover, or offshore to a depth of 20 to 85 meters.

These differences, which help to make the Portuguese a lot easier for humans to deal with, presumably have a bearing on its taste in the human mouth too. It is coarser and tougher all round; its diet is somewhat different, from its greater tolerance of mud, and whether or not its eating habits make it grosser to eat, it does go about its own feeding more grossly. There is something admirable about this, though; if we had to work as hard for our food as any of them we wouldn't have gone to all that trouble to get mutated the last time or two. (It will be objected that animal food is much harder to come by than plankton. But here and there a few of us ought to cling secretly to the pathetic fallacy. It has had its uses, after all.) In mitigation of the Portuguese, it is really not as uncouth as it sounds; it has a better sorting mechanism that the *plate*, so can afford to be less particular about what it takes in.

The oyster eats, as it breathes, by pumping water in and out of its valves more or less continually, at a rate that is awesome. At the most favorable temperature, said to be 77° F. for at least some species, the *plate* takes through itself one liter an hour, the virginica 3 liters, and the Portuguese 5 to 6. At this last rate, an average-sized human would take in the contents of sixty-two full bathtubs every hour, or in twenty-four hours would work through a large public swimming pool. In the case of the oyster, the pumping mechanism for this vast operation is nothing but a lot of little lashes, or cilia, lining the pleats of its breathing apparatus, or branchiae; said lashes being in a frantic state of motion that creates a current, whereby the sea water and the particles of nourishment in it go roaring through the valves. As it takes at least a million oysters to make a healthy natural bank, and each one is picking out whatever it can eat before expelling the water, it all adds up to quite a commotion and quite a job of filtering. To be strictly truthful, since the microscope magnifies distance but not time, the process only *looks* all that frenetic. It seems calm enough to the oysters themselves.

They are said to be omnivorous, eating anything available in the plankton line, all sorts of animal and vegetable specks, debris of organic matter, eggs and larvae in season, with big differences according to habitat in what is available. By their location in or somewhere near the teeming tidal muds of the earth the Crassostrea would seem to be the gluttons of the tribe. Way out at the deep edges of the continental plateau, where the green plant things or diatoms thin out and finally cease for lack of sunlight, the Pycnodonta have to be more carnivorous, but they don't concern us and will not be mentioned again. Our Ostrea, when living where it chooses, appears to have a well-balanced diet, including the common-garden tidbits that are among the prettiest and most thought-provoking of all the 800,000-odd forms of life in the sea. These are the various kinds of protozoa, the little single-cell creatures of infinite grace of design and a nomenclature more ponderous than whales, on whose flimsy filaments rests the whole weight of the second most solemn question there is. In them, animal and vegetable life converge, and legend dies for the human mind; or if you want to place that in time instead of in the scale of life, let's say it dates with the invention of the microscope, which amounts to the same. Anyway, these portentous morsels, measurable in thousands of a millimeter, lovelier than snowflakes, with genius in the plan of each and madness in their variety, are in their daily lives fodder for the *plate*, as for any other animals the right size to eat them.

The general range of the oyster, in latitude, is determined by temperature. It seems surprisingly wide in the case of the *plate*, considering how little it takes to kill the young or keep the adults from procreating. Leaving out two Ostrea in other parts of the world that go by other names and may or may not be of the same species, one on the south coasts of Australia and New Zealand and one that is cultivated in a small way in Brazil and Argentina, the European *plate* extends from Sebastopol on the Black Sea to latitude 65 on the shores of Norway. In southern Norway it used to be very plentiful. It has been seen in Iceland, and off Cape Guir near Agadir on the Atlantic coast of Morocco, evidently its extreme southern limit. It does fairly well here and there on the nearly tideless Mediterranean, though raising a question there as to whether it is actually of the same species. It certainly doesn't taste the same to an ordinary palate, but then aside from difference in environment, it doesn't usually get on the Mediterranean anything like the élevage of the best Atlantic oysters.

With such a spread of coastline possible to it, you might think it could be bred in any number of places. That it can not, and that not even the whole Morbihan but only one very small piece of it should have a near-monopoly in that part of the business, is due to a combination of good, or let's say ineluctable, reasons.

For a long time, when nature was still running the show and in spite of certain lapses on its part, the *plate* did proliferate over a great part of its range. The Romans found them by the ton in England and elsewhere, most excitingly at the mouth of the Gironde; from late prehistoric times, especially in Denmark, there exist the huge man-made mounds of shells called kitchen middens, a polite term for the more rewarding sort of garbage dump. If Charlemagne's armies had had the taste for oysters that Caesar's did, which is doubtful, they would still have found plenty of places to fill up on them; and many of those stretches of coast apparently went on being just as rich up to the early 19th Century. Not quite all. Gradual changes of shoreline did them in in some places, often smothering them in river deposits that have pushed the coast out several miles since Caesar's time or even within living memory, as at one point off Brouage, near Marennes, where the *plate* prospered not so long ago and would now be five feet deep in mud. The above-mentioned garbage dumps indicate that the now less saline Baltic Sea used to be salty enough for it, when the opening to the North Sea was wider and deeper.

This is all in reference to natural banks, that is to conditions that the oyster would find congenial enough, in temperature, food, bottom and so on, so that it could perpetuate itself, by itself. Once the banks were mostly destroyed, and the big agent in that was not natural developments but man, the picture was altogether changed. Places famous for oysters when the oyster had been out deep and on its own, and that barring human misbehavior might still be excellent for the adults of the breed, turned out to be no good at all for the *captage du naissain*, by the use of collectors, or for rearing the very young.

A big displacement of the oyster was involved to begin with, from regions that would rarely if ever uncover, up to the intertidal strip, where the collectors could be handled and the parks tended at low tide. There had to be natural banks nearby to provide spawn, but in combination with a set of inshore conditions so exacting, an electronic brain would go berserk computing them. Every need of the oyster takes on another complexion when it is out of water part of the day; every hazard to the adult is hugely intensified for the young; furthermore, as the Greeks knew but many people later had to discover all over again, any given generation of oysters may grow splendidly in a place where they give forth little or no genital products, or where most of their larvae die.

A lot of places were tried, and a lot of people were ruined. In some places, an untypical turn of temperature or rainfall might create the right conditions for a year or two, or even a run of eight or ten years, opening the field to disaster afterwards. A few, including Marennes and parts of the Ile de Ré, worked out well and continue to breed a certain number of *plates* along with their great quantities of Portuguese. But these have never had the quality of the oyster of Morbihan origin. Even in the early days of élevage, oystermen in that region preferred to get their "eighteen-months" all the way from Auray when they could, and found that it paid in spite of the great trouble and risk of transportation by sailboat; within a year the Breton oyster would have grown nearly twice as fast as the local one. The Gironde has its own local pride, which forbids this fact being mentioned very often, so we will drop it. The indisputable fact is that the Locmariaquer neighborhood became what it is because no other bit of geography has been found as well suited, on as large a scale and as much of the time, to be a nursery for this particular infant.

The stately procession of barges with their white cargo in the dawn is unique; you won't see anything that begins to match it anywhere else; and seen from the shore, there is such quiet glory to the sight, they all seem headed so serenely for regions beyond the sun, for a while as they pass you can feel man and the mollusc alike freed of their origins in time, and imagine that whatever that whiteness is, it has been happening that way forever. Actually it is quite recent. The whole business of the tiles dates only from about 1860.

THERE was a girl in Locmariaquer who wanted to get out. Her name was Marie-Yvette, with her mother's last name, but she came to know her father by sight. He was the proprietor of one of the better restaurants in Auray. Probably he had seen the girl's mother around the little oysteryard his wife managed on the side, and he may have been briefly amused by her deformity, or by something fine-grained in her mind and spirit out of the peasant run, out of his own run too for that matter, or her being a forty-odd-year-old virgin; and perhaps the thing was not as grotesque as it came to seem later. She was not so lame before the birth of the child, which she almost didn't survive, and even after everything had happened there remained a glimmer now and then of some strange shy rapture in her smile, hard to account for. She couldn't read or write, never having been to school at all, but was very quick at figures.

Marie-Yvette grew up fairly happily through her First Communion, a wonderful day. She and her friend Yvette, without Marie, had it together and nobody could decide which was prettier or said anything, at least not in her hearing, about her not having a father. The two of them held hands and giggled at the customary remarks about the little brides, and her uncles who were home for the day, both jolly boule-playing types although one had never married and the other had lost his wife and child in a fire, got drunk and gave them some ersatz candies they had picked up somehow. Her mother and grandmother were in their best coiffes and very proud, the mother gently incredulous in her joy and the grandmother more astonishingly, for sorrows and hard work had turned her to a mountain of granite, singing. Long ago she had lost two other children, her husband had gone down with a sardine boat, her wastrel brother had sold off the home they grew up in; when her only other grandchild was brought dead out of the fire, people said that some weird ancient imprecation had come from her mouth in a speech nobody could understand. But this day with the bagpipes playing, straight as an empress like everyone else her age in her massive black under the stiff spotless coiffe, she laughed at her younger son's fooleries and gossiped and sang all the words of old ballads that she knew better than anyone. Yvette, without the Marie, made eyes at one of the Germans and got slapped for it and her handsome older brother made eyes at Marie-Yvette, for the first time. The town was packed with refugees from Lorient and Hennebont who were always crossing themselves, the women especially, at the sounds of bombing up the coast but nevertheless stared appreciatively at the little girls all in white, almost as if some charm against danger had entered into them for that day. They felt themselves that it had, for that day and a long time to come. It was the spring of 1944.

One of the uncles was picked up that night for deportation to Germany and was shot on the way; the other was killed three weeks later in Lorient. Until then the family had kept a half-title to the last of the two big farmhouses they had owned; now Marie-Yvette and her mother and grandmother moved to the hut by the shore.

Five years later, when the girls were fourteen, Yvette went into the oysters, but Marie-Yvette would not. Her mother had done that until she was too crippled and her grandmother until she was over seventy, and the girl wanted no part of it. She wouldn't even stop to talk to Yvette's brother who loved her and had a job in the best chantier around. She wanted to go to school some more and become a secretary and move in the world beyond misery, where her father lived; and somehow her mother, who would have cut off both hands for her, managed it. She took any hard or filthy job anyone would give her, letting herself be ordered about like an ox. She even found an oysterman who would take her back at reduced pay and was out in the cold all day for two winters, never speaking of the pain in her hip, or of her daughter's shame and almost disgust at the sight of her, which the girl tried not to show. The three of them lived mostly on bread and sea snails in order to pay for the school and bus fares to Auray, where Marie-Yvette saw her father prospering more than ever after the war and never bothering to look at her when they passed in the street.

There were times when the humiliation lifted and she longed to do something grand for her mother, and would bring in wild flowers and sing and dance around the room, or sit peacefully with her head in her mother's lap, until a fit of anger would take her at her mother's timidity in either caressing or scolding her. What discipline she had known outside of school had had to come from the grandmother; to the mother, the sound of the word Maman had never stopped being wonderful. She could never quite believe that anything so well-formed and flashing was really of her flesh.

Marie-Yvette got a job in Auray at last, bought her mother a new black cotton dress which of course was the only kind she would wear, and one day when she was eighteen came home with a bottle of muriatic acid and drank it. It took her half the afternoon to die, on the bed she had always shared with her grandmother, while Yvette with her baby and Yvette's brother and the others waited outside. They had heard her screaming way over at Saint-Pierre.

If a count could be made, it would probably turn out that at least half the references in English literature to oysters, a very common article of food in Great Britain until recently, have been based on what was a far-out exoticism before the 18th Century and since Linnaeus et al. has been an error of nomenclature. Nothing shows more clearly the basic romanticism of the English-speaking peoples. "He is a pearl within an oyster shell"

(Shelley). "Rich honesty dwells like a miser, sir, in a poor house; as your pearl in your foul oyster" (*As You Like It:* foul in this instance meaning not nasty-tasting but outwardly unprepossessing). "The firm Roman to great Egypt sends this treasure of an oyster" (*Antony and Cleopatra*). "Then love was the pearl of his oyster" (Swinburne). It might even be as good a means as any of classifying English letters, with Shakespeare of course in all categories. Wordsworth, for instance, seems never to have mentioned oysters at all; the fact speaks volumes. And what could give you a more striking clue to Browning's vision than his having mentioned them repeatedly and *never* associated them with pearls? "Than a too-long opened oyster"; "And laying down a rival oyster-bed"; "Turn round: La Roche, to right, where oysters thrive . . ." It took real intellectual stature to think of them like that, considering the set of the centuries before; it is rather grand. And popular association still hasn't caught up with Browning, at least in America; outside the cookbook, oyster still means pearl to most people.

There are some deep reasons for this, beyond the obvious one of cultural lag, as it used to be called. Actually in our present conditions of life the thing without price would seem to be more the oyster, if it were a good one, than the pearl, the quality of fake jewelry having risen in exactly inverse proportion to that of real seafood. But for ages in England the word, in reference to the very best too, had the weight of "a dime a dozen," or more literally something like a dime a barrel. "But thilke text heeld he nat worth an oystre." Chaucer could even speak of "oystres and mussels and oothir swich," as if they were all in cans together in the supermarket. The romantic fallacy, the pearl idea, seems to have come in with the great voyages to the Orient, where the mollusc that came to be called the Pearl Oyster lives.

The true name of the creature is Meleagrina margaritifera. It is of the family Aviculidae and is related to the mussel, not to the true oyster. For one thing it lies on its right side, something that would cause any proper oyster to be ostracized (which does not mean turned into an oyster but discriminated against by ballots made of broken crockery). It is true that any bivalve, if annoyed by the intrusion of a grain of sand, a worm larva or other inconvenience, may build up around it an object resembling a pearl, but if made by an oyster it will not be very pretty. The mineralogy is wrong; genuine pearl substance is aragonite, and in the oyster shell calcium carbonate takes the form of calcite throughout. Margaritifera, which is the best of the pearl-producers but by no means the only one, hasn't even much outward resemblance to any of our familiar oysters. But then there are living species of oysters that are a far cry from what we see on the plate too, some with a wild jagged crenelation of shell so deep you marvel at the perfection of fit of the two valves and the exigencies of their continual opening and closing all through life; if you hinge your two hands together at the base of the palm and try to make a perfectly closed dome out of your ten fingers you get some idea of the challenge. Incidentally, the adductor muscle does the closing; opening is performed by an adjacent ligament made of horny conchiolin, common to all bivalves but of a unique triangular shape in the oyster, and which is an integral part of the shell.

A word on the second most widespread association with the topic—the months with *r*. Bartlett's *Familiar Quotations* gives its first literary appearance in 1599, in William Butler's *Dyet's Dry Dinner:* "It is unseasonable and unwholesome in all months that have not an *R* in their name to eat an oyster." The idea derived no doubt from the difficulties of keeping oysters fresh during transportation in hot weather; it cannot have been out of respect for the spawning season, since the supply for a long time after that writing was considered unlimited. There is nothing wrong with an oviparous oyster at any time, and an individual of the other kind will be unpleasant only during the eight days of incubation. Otherwise, the only thing against them during the *r*-less months is that the stress of procreation makes them rather thin and watery.

Oddly enough the most famous oyster metaphor in all literature, as well as the most mysterious, touching most delicately the downward spiral of significance of this above all sea creatures for the human mind, has on the face of it nothing to do with a pearl. Or perhaps it has, but we would rather think not. "The world's mine oyster, which I with sword will open."

IT would be delightful to come over by helicopter on the morning when the first tiles go out, a little before high tide.

You might start from Vannes, where Caesar defeated the

Veneti, and zigzag up the Gulf, the beautiful Little Sea the region is named for, with its twisting indeterminate fingers pushing among groves of holly and mimosa and once sacred oak, and overgrown geometry of old salt basins back toward the marshes here and there, and crazy whirl of currents around some of the islands, said to number one for every day in the year but of which there are really about half that many. Some may have disappeared in the last millennium or two, from human iniquity or other causes. Come down low and see the wild skeletons of fishing boats beached wherever the sea chose, all eyes and warnings in the fading dark; go higher and see the three-thousand-and-some giant rocks upended and aligned out of some overwhelming prerogative, to do honor with blood and horrible shoutings to that deity; the morning dew makes the crosses scattered among them prevail again, up to a point.

It was a terrible job taming the old presences in this land, and perhaps it will never be finished. Long after the menhir people had vanished, whoever they were, magicians roamed the woods, and the dolmens, having roofs and an association with skeletons, became the favorite homes of the dreadful dwarves of the Celts. The more dreadful korrigans combed their long blond hair beside the sacred springs, and nobody knows where old Merlin went to die. No drink was as bracing to a Breton warrior as stolen wine, it had to be stolen since they had none then either, mixed with the blood of a Frank.

But just as topography, unless the sun is up and strikes the roiling currents or you get a glimpse of the razor-blade cliffs of the Côte Sauvage, way off on the ocean side of the Quiberon Peninsula, the only wizardry of the scene in mild weather is a balm of peace. There is something hypnotic about the Gulf in that respect; some tender hand or breath must control its character; it is all a cradle, moving, bemused, strung with the baubles of summer villas but discreetly, under a blowing baby-laundry of fresh little clouds; the port of Vannes itself is a place for dreams to moor and rock gently through the night. Looking down on it, you would not suppose people had ever been in battle there, or quarreled over the price of anything. Up at the opposite end of the Gulf, some twelve or fourteen miles away in flight distance, the high white lighthouse of Port-Navalo pins all the gauze and laces of shoreline together, and that is where the bells ring loudest, cheerfully more often than not, although some of the more muted bells that echo those seem to ring from the drowned and haunted villages that may be there; nobody would swear they are not, or that they won't some day be resurrected, with their inhabitants.

However, the thing to keep in mind is that billions and trillions of oyster eggs at that very moment are either being shot forth from their mothers' envelopes, or have already turned into larvae, to be eaten by practically any other creature in the neighborhood or swept to some other death. At most a hundred out of a million, with luck, may end up on the tiles.

The barges back in the Gulf have to start earliest. You see the first one edging slowly out, a white solemn rectangle but leaving no fire or music on the beach behind; if there are any children there they are still asleep; everything is perfectly still. A second starts moving out from one of the islands, and suddenly both are seized by the same convulsion, like a last fit of resistance to this journey, and start spinning and straining against their ropes, but the pilots are used to that and soon have them following quietly. By now there will be half a dozen moving the same way, and then so many you can't tell where they came from except by the far bigger stacks, not yet the abstraction of whiteness they will be in full daylight, left on shore for another day. More are pushing up from Le Bono, down at the second embranchment of the River of Auray, and as the two processions begin to merge at the mouth, all the barges of Locmariaquer, of all possible sizes, drawn by big boats and small and at the modest end of the scale propelled by an outboard on the barge itself, come sweeping out to join them, and the long line will start battling out through the straits, for they must be where they are going when the tide turns and at this point the passage, only a kilometer wide, is in turmoil.

It may still be too early for the bells, but half the town's population is out there so everyone has been awake for a long time. An air of elation has taken over; here and there someone shouts a last joke or admonition from the shore; there are figures waving, leaning out windows, running in the street as if everyone had something exciting to tell someone else. Later the barges will begin scattering every which way and at some places will meet a contrary traffic of the same kind, coming out where these are going in, along with some barges that are unloaded where they were, without any journey, but here after the narrow throat of both river and gulf they all have to turn to the right around the point—there is nothing doing on the Port-Navalo side—and proceed in state together at least to the River of Crach. The sun rises around four and if it is that late or a little later and not misty, glory breaks over them from the sea and the sky together as they turn out into what is still not open ocean, except by comparison with the Gulf, but merely the wide crescent Bay of Quiberon.

At the right height you would see the confluence of waters and all the holes and tatters of coastline responsible for the business. There are few trees of any size, but in one place there is a surprising alley of oaks planted ages ago between two erstwhile farmhouses that have been stuccoed and are both shut and dead-looking; the one with green shutters, nearer the shore, was bought a few years ago by a Parisian family, who will not be there for another month. The tide is up almost to the doors of the pink one, and if you came low over the dyke alongside you would see the four slim-necked horses tinged with coral and gold, thrown into a paralysis of watchfulness by the roar of motors from the sea. Or perhaps the white fleet moving into the sunrise is what they are exiled from.

The farmers haven't appeared yet; the others left by moonlight to go out with the tiles. But you might catch a glimpse of old Françoise out among the little scrub-oaks beyond the hut, at the very edge of the sea, with her walking stick. It is the same every year. When the noise from the point wakes her on that day, she can't resist coming out to watch, and for a few minutes the ineradicable anguish is washed from her face. She is young, and as beautiful as the horses, and her daughter moves in radiance with her across the bay.

THREE

THE HUMAN RACE is said to be growing taller, but its stomach is evidently shrinking. In the past, where oysters were eaten at all, a plate of six or a dozen would have been considered ridiculous. Another important fact of mollusc-lore is that although oysters have had certain ups and downs in popularity as food in some parts of the world, in general man seems to have been eating them, if not gorging on them, from the minute he appeared. Or since he was kicked out of the garden, since Adam is pictured as a vegetarian before that, and there is no mention of any seacoast nearby. Anyhow, if you could bring together the shells of all the oysters consumed by humans since we have been around they would probably stack up to something like the Himalayas and then some, and this has mostly not been a matter of word getting around from one area to another. Oyster-eating was world-wide long before the Phoenicians and the Vikings. Apparently people took to it independently, while still in the grunting stage, on all the temperate-zone coasts of the world where the oyster of any sort was available.

If that is a slight exaggeration, it is not a gross one. Nobody would say that the inside of an oyster is very pretty to look at, and there must here and there have been tabus against it among others than the Jews, as there may have been some fetishism for it, because of its association from way back, correct or not, with virility. At any rate, the famous kitchen middens are evidence of a vast and widespread consumption of this article of food by prehistoric man. The "bold man who first eat an oyster," as Jonathan Swift dourly called him, has to be imagined opening it with a rock or a club or perhaps by hurling it over a cliff seagull-fashion. Anyway some of the mounds he and his friends left, in which the shells are mixed with bones of birds, mammals and fish, are so huge, they are supposed to have taken centuries to accumulate. Others, coming down closer, contain bits of pottery and worked metal. Some of them, as in Senegal, at Lake Diana in Corsica, and at Saint-Michel-en-l'Herm in Vendée, have been the subject of a great deal of speculation, some students of such matters maintaining that although they include signs of human industry, they are not man-made but the remains of huge natural oyster banks thrown up by an upheaval, slow or sudden, of the soil. The more recent of these shell mountains may be the site of some big business for the exportation of oysters, smoked or in brine. And there is a view we won't pursue, that by analogy to the ancient animal-shaped earthworks in the United States, takes one or another of them for deliberate representations of sacred dragons, serpents, etc.

Whatever they are, it took a lot of oysters to make them. Oddly enough the Morbihan with all its prehistoric monuments, and whose shores were presumably teeming with oysters throughout antiquity, seems to have no conspicuous dump or deposit of the kind. Perhaps its early inhabitants didn't like them, or perhaps they had the sense to throw the shells back into the sea, unlike most fishermen a few thousand years later.

Among others, the Chinese have eaten and cultivated oysters for several millennia. The coast Indians of North America relied heavily on them. The famous "fishweir" uncovered under the heart of Boston not so long ago, consisting of a strange arrangement of wattles surrounded by remains of a huge oyster bed and considered to be between 2,500 and 3,700 years old, may perhaps have been not a weir at all but a collector for oyster seed. (T. C. Nelson in "The Boylston Street Fishweir," 1942) It clearly did serve that purpose along with any other it may have had, and if the Indians hadn't wanted the oysters they would presumably have put their wattles somewhere else. In Maine one of their mounds of shells is estimated at seven million bushels. There was a similar mound, in France, near Lyon, the old Lugdunum, but then the Romans were as exorbitant about oysters as in everything else. Aside from their own production of them near Naples, they had them sent to Rome, and to the rich provincial capitals as well, from Bordeaux, Britain, Africa, the Black Sea, everywhere. Anybody rich enough ate them at the beginning of every meal and a few may have vomited so as to be able to eat some more, and there were probably many socialites like Nero who boasted of knowing the origin of any oyster by its taste. "The palm and pleasure of the table," Pliny called them. How they got there fresh is a question; probably quite often they didn't, unless they could be dipped in sea water now and then along the way, and no doubt the repulsive-sounding sauce, the garum, they were often eaten with came about as an antidote to any slight putrefaction, being itself a super-putrefied brew made from the intestines of other sea creatures. Of the several varieties of garum, the most prized was from Cartagena in Spain. It was the most expensive liquid in the world, had a bitter taste and a horrible smell, and was wildly exciting to the appetite, at least of those jaded eaters. An inferior version was still being sold in fish stores in Constantinople not so long ago.

The garum was invented by the Greeks, according to Pliny, and is probably as good a single explanation as any for the fall of both Greece and Rome. Seneca, without reneging on his love of oysters, speaks of this way of serving them as a cause of the bad health of the rich in his time.

Leaving out the Orient, there was to be no such traffic in shellfish again until the 19th Century. On its own terms the oyster was flourishing. The Atlantic banks of Ostrea edulis formed, if not a solid line at least a fairly plentiful string from Spain to Scandinavia more or less, and there was never a time when they were not exploited, where accessible, and to the point of being a natural metaphor. "Faut-il endurer ce sanglot / Ainsi comme huîtres de Quancalles?—Must we suffer this tribulation, like Cancale oysters?" (From the "Mystery of the Siege of Orleans," 15th C.) A basic French thought on the subject, to be seen in Boileau out of La Fontaine, and in crisp contrast to the romantic English idea, is that "the oyster is for the judge, the shell for the plaintiffs." Only Alexander Pope took that up across the Channel, which may or may not be a tribute to the British legal system. Generally when not fixed on the fallacious pearl, the English have been more esoteric in their deeper feeling about this particular mollusc. "An oyster may be crossed in love . . ." "Then love may transform me to an oyster, but I'll take my oath . . ." Obviously nobody would think that way about a scallop, even aside from their respective economic positions, but the abundance of oysters at the time did have something to do with it. There is an occasional French lyrical touch too, as from poor Cyrano de Bergerac, a playwright remembered only as a character by Rostand. We will herewith restore him to his proper lights. "You have never seen the sea," he wrote, "but in an oyster on the shell."

The oyster on the shell, *l'huître à l'écaille*, is in distinction to *l'huître sur paille*, taken from the shell at its point of origin and packed in straw for transportation, mainly to Paris. This was done on a considerable scale in the 18th Century. But generally speaking, between the collapse of the Roman system of transportation and the development of the railroads, supply and demand went together as usual outside of Russia; in most places there was very little of either. Inland the oyster was mostly for royal courts; on the coast it was the food of the poor along with everybody else.

The most famous oysters remained those of the Bordeaux region, as in Caesar's time. The poet Ausonius, who lived around there in the 4th Century, speaks of them, with the same enthusiasm as Montaigne, who visited Bordeaux on a parliamentary mission in 1581. "They brought us oysters in baskets," Montaigne wrote, describing a shore dinner in the neighborhood. "They are so agreeable, and of so high an order of taste, that it is like smelling violets to eat them; moreover they are so healthy, a valet gobbled up more than a hundred without any disturbance." And in another passage: "To be subject to the colic, and be deprived of oysters, is two evils for one; since we have to choose between the two, we might as well risk something in the pursuit of pleasure." He was indulging in a little of his hypochondria in that sentence. There are only two or three other references in all literature to healthy oysters, without sauces, giving anybody a stomach-ache, and Montaigne himself tells the following: "A doctor, a great eater, who was killed by an attack of apoplexy in front of a paté de foie gras, had been able to absorb thirty to forty dozen oysters with impunity."

The figures he gives on individual consumption are nothing out of the ordinary. Casanova ate fifty every evening with his punch, so he says, as a man in his fix nowadays, faced with the grind of perpetual romance, would probably take pep drugs, but others did as much out of nothing but a healthy appetite and enjoyment of their food. "I have just come from Le Havre," runs an 18th Century document, "where I was summoned by the news of the death of my uncle, whom I found at breakfast, eating his eighth dozen of oysters."

And here is Brillat-Savarin on the subject:

"In 1798 I was at Versailles, in a post as commissioner of the Directory, and was in rather frequent contact with Monsieur Laperte, court recorder of the department. He was a lover of oysters, and complained of never having eaten his fill of them, or as he put it, having stuffed on them. I decided to procure him this satisfaction, and to that effect, invited him to dine with me the next day. He came. I kept him company to the third dozen, after which I let him go on alone. So he went on to the thirty-second, that is for more than an hour, for the maid opening them was not very skillful.

"Meanwhile I had nothing to do, and as it is at table that that is really distressing, I stopped my guest just when he was going strongest. 'My friend,' I said, 'It is not your fate to stuff on oysters today, let us dine.' We dined, and he behaved with the vigor and bearing of a man only then starting to eat."

Since the accounts are nearly unanimous in showing the multi-dozen eater quite sound in digestion afterwards, one has to wonder how many Henri IV took on, to get his famous stomach-ache from them in 1603, that being one of the rare references in history to any such malaise. He may have been sick beforehand, or it may have been a complication due to the drop of arsenic he was given at birth. In general he had the reputation of eating not dozens but hundreds at a whack, which probably would have given him his name as a *vert galant* even if he had done nothing else about it. A number of French medical works, beginning with one published in 1689, praised the properties of the oyster, so the empirical evidence must have been favorable.

In the American colonies, opinions on the value of oysters were erratic. The New England settlers had learned from the Indians, as well as from their reading of the classics no doubt, to regard them as either delicious or good nourishing vittles, as the cultural case might be. On the other hand the people of Maryland, in a suit of 1680, listed among their grievances "that their supply of provisions becoming exhausted, it was necessary for them, in order to keep from starvation, to eat the oysters taken from along their shores."

But then came the explosion. The sleeping giant, later known as the Consumer, awoke with a yell, and the cry for oysters and more oysters was heard across every civilized land. Rome had been nothing to this. Shipyards sprang up, coast after coast was raked to fill the insatiable maw. In France a royal edict of 1681 had imposed strict regulations on mussel fishing but said nothing about oysters, for the reason given by one of the king's agents in the Admiralty, that "the natural banks are inexhaustible." True if the social structure had stayed put. Another royal edict of 1726, restricting other kinds of fishing, specified that oyster fishing would continue by means of the iron drag "in the same manner as practiced heretofore." Scholars will decide whether the ostreomania that soon after swept the world came about as a result of the French Revolution (nouveaux riches) or the industrial revolution (railroads, conspicuous consumption) or a Napoleonic fad, or a wave of general gourmandise, or our first envoy, Benjamin Franklin, landing in Auray in December 1776 and quite possibly being given oysters by the welcoming committee in the hope that he would advertise the local product in Versailles and Philadelphia.

However it was, by the middle of the 19th Century the natural banks were close to exhausted in the Morbihan as in Maryland and in most other places where there had been any commercial exploitation of the oyster, of whatever species. It was on the way to following the great auk and the passenger pigeon, through the same human agencies. Hence the bitter tears over its story in *Alice in Wonderland.*

THE few TV sets in Locmariaquer are in houses of the *gros richards;* there is none in any ordinary family's house, nor in any public place; the programs are pretty terrible anyway. There are, however, 41 telephones, distributed as follows: 2 at the two little hotels that haven't much to do nine months of the year but are packed in summer; 1, baker; 1, café; 2, the two butchers; 1, La Mairie, including Police, Marriage License Bureau, the public school which is in the back yard, etc.; a small assortment of well-off citizens; and 22, oystermen, of whom the two big shots have two phones each, one at the yard. The church has none, neither have the real estate office, the garage-*cum*-bus terminal, the convent with its big school plus nun-nurse for all the countryside, who does however have a tiny car; nor the Catholic boys' school directed and taught by the ubiquitous Abbé, who also has a midget car for emergencies, as well as a bicycle for exercise and to save gas. The really crucial phone, more than the one at the Mairie which after all is really important only about once a week when somebody is killed in a car or motorbike crash down the road, is the one in the booth at the Post Office.

This bit of the awesome national bureaucracy is housed downstairs in a little faded blue square building set back from the road, approached by a dirt path between two somewhat unkempt but multitudinous and sprightly patches of flowers. The Postmistress, in her fifties and with a head to be seen around the portals of a thousand cathedrals, those that were not hacked off during the Revolution, presides over the drab interior with the charm and dignity of great hostesses, born to a long tradition of high diplomacy. Her beauty lends itself to this, once you get the hang of it, past rimless spectacles, ashy blond hair wisping away over narrow eyes and forehead like a teacup in flight, a nose to cut hedges with and resemblances to the classic schoolmistress, in gait and prominence of knuckles and the kind of nervous efficiency that in any female authority, over anything, warns you of squalls ahead. An error in this case. Only the top layer crackles; all is kind and true beyond. Her opposite numbers in almost any larger town, such as nearby Auray, would be flabbergasted; they are closer to the heart of bureaucracy and have a coat of arms consisting of the Refusal Net on a Field of Unconcern, not that all quite manage to live up to it but the inspiration is there. The noblesse of our Postmistress is of a different order altogether; except in the life of the village her office is so unimportant she is not even required to be rude, and it happens that her nature doesn't require it either. She sits as she stands, straight as a lily in stone, or a live flamingo, the quick head bobbing courteously before every deep revelation and 25-centime stamp.

Birdlike, she closes up at noon and rests from the flutish warble of her working hours, for in spite of everything she is not a mere individual, however superior, but one of a breed, which may chat or simply utter in the family but must do its business in song. That is how you can tell a foreigner or a person of low degree, or else the highest; they *say* Thank you and Goodbye. She has one musical phrase for "Au revoir, Madame," and another for "Mais pas du tout, Monsieur," and one for selling national lottery tickets to those who can afford it and another, less chirrupy, for those who can't, and a gamut from Mozart to Schoenberg, depending on how long she has been kept waiting, for "This is the cabin in Locmariaquer, calling the number 27 in Auray . . ." or whatever it is. The booth is in use quite a bit and she must personally contrive each connection, while carrying on the postal business, under which is included the National Box for Savings; meanwhile also, in tones of ordinary human speech, that is without the slightest facial or vocal strain more or less talking and singing at the same time, quietly giving directions of one sort or another to the dark-eyed young girl apprentice who shares her cage and keeping control over the small grandchild or two who are usually in there too, most often on the verge of eating several thousand francs' worth of stamps.

All of which notwithstanding, and although it is nearly closing time, and although furthermore she has just lost a good twenty minutes on the phone finding out how to charge Monsieur for a cable addressed to Chapel Hill, North Carolina, not omitting to instruct José the apprentice when some upper echelon of the system does find the right manual to consult for this novel contingency ("Take your notes, José. Vous voyez? Compound foreign names are charged as one word. Only the big offices have this information; it is not supplied to us"), she serenely undertakes to help Madame make a phone call no less challenging to Vannes, the department capital. In fact it turns out it would have been easier to phone the Black Hills of South Dakota; to begin with the number does not exist.

Madame is trying to call the Society of Many Learnings, to ask when it is open, having been told that only in the library of said Society will she find books on oysters. Books by oysters, it might as well have been. If it is not the Mutual Assistance Society of Lay and Clerical School Teachers, it must be the Morbihan Society of Ornithology and Natural History, unless it is a branch of the catch-all if not godlike Society of General Enterprises. A whole opera ensues by phone, with some of the same cast that have just performed in "Cable Charges for Compound Foreign Names." The confidante role, sung by the chief operator in Auray, rates a proper name, Madame something or other; she and our Postmistress evidently went to school together. The one in Vannes is addressed simply as Vannes, in a different register.

The plot thickens, the characters multiply, the clock ticks toward six; the little children eat their fill of stamps, while in the course of a terrific trio, somebody is reached at a certain café who has some connection with a society that ought to know something about the Society of Many Learnings. In fact a clue is provided, involving however a toll charge. "I must verify," sings our soprano, still dulcet but no pushover; she can play Medea too in a pinch; "I cannot charge my client"—oh lovely aria—"for an uncertainty." She cracks only once, having had to explain something three times to faceless Vannes, the shrouded, insinuating one that no libretto can be without. Many in the story have been roused to passionate effort by the mere name of the society in question, for although Bretons they also belong to that extent to the nation of Culture and the Great Exam. Not Vannes; many learnings or none are the same to her. She whispers something, not to be heard by the audience; whatever it is the Postmistress, who has been pushing back damp strands of hair and had come close to lapsing into recitative, is instantly piqued into producing not only a smile for queens to envy, through lips that could do the work of scissors ordinarily, but also the absolutely finest cadenza of the season. "Mais non," she sings, smiling into the telephone, and all who hear are breathless, even the children are stopped by the sweet purity of the notes, as lushly tumbling as the nightingale's, "mais non, je ne suis pas irritée, mais pas du tout . . ."

That fixes Vannes. She slinks off the wire, leaving the cabin at last in communication, not with the Society of Many Learnings, which has no phone, nor with the home of the director of the Society, who has none either, but with a character out of a foreign-language textbook—the wife of the brother of the director of the Society, or as we would say in English, the director of the Society's brother's wife. No, that can't be right. Anyway, the Postmistress and her client have agreed that charges can reasonably be assumed this far down the scale of uncertainty, and are rewarded by a flawless recitation from page 3 of the Exercise Section. Yes, Madame, the Society is open every day from nine to twelve and from fourteen to eighteen o'clock, including Saturdays. Splendid; thanks are chanted all round. Now that the goal is around the corner Madame would not be so tactless as to explain what she is after. The very least they must be assuming is that she is a distinguished scholar of Prehistory, a favorite field in these parts. Alas that it is not so.

The scene changes; we have to leave our central figure to return to the niche on the cathedral, with the lyre and the telephone as her symbols and the moving legend, "Pas Irritée."

The Society of Many Learnings is not only shut tight as a tomb but is not even supposed to be open, ever. The concierge proposes instead a visit to the Museum of Prehistory in the same building. Furthermore, on a corner of a building across the street are a pair of medieval grotesques that are called quaint in the guidebook but furnish a shocking note in this drama, considering what has gone before. They are called "Vannes and His Wife"; it is worse than the sex of the oyster. On top of that it develops that twelve of the fourteen tenants at the director's home address, all with identical doors opening on the resounding cavern of a stairwell, have the same name as the director. A baker's dozen of sisters-in-law get involved but at last he is found, gloomily presiding over a closing-out sale in his store down the street, or what was a store. In the dim light from the door all that seems to be left is a few ancient bottles of laxative and hair tonic, a dusty pocket-comb or two, a couple of comics. No place on the cathedrals for him; too good to be a postal clerk, too unlucky to land in one of the few professors' chairs available, inept at business, with black-rimmed glasses and a slew of children to support, he is the damned and forgotten man of all Europe, the provincial intellectual, in this case a Doctor of Prehistory.

Madame's heart sinks. All she can possibly do is to add insult, from the baseness of her request, to injury, from belonging to a country in which Ph.D.'s are so well paid that a million or so can afford to swarm all over France every summer. In fact a glaze, not of contempt but as though the word had revived too long a series of tragic disappointments, comes over the director's face at the mention of oysters. He is tall, about forty, and built like an expert skin diver except for the worry and dilapidation; he speaks and gestures with frantic rapidity, beating invisible gargoyles back into the dark corners of the store. He is polite, however, and helpful, naming a certain book on oysters of the Morbihan, written by a local sea captain. There is of course a copy in the library of the Society of Many Learnings, but unfortunately the Society and its library are private; the book will no doubt be found in the Municipal Library, at the Mairie. Vannes and his wife indeed; Vannes is just the bitch she was on the telephone, taking cover in the name of a good and lovely town and using this desperate scholar as her mouthpiece; only Madame has no retort to sing. She has been moving up to now in a magic of favors from all quarters, rich and poor and including the government office concerned with oysters, but that is just the point and it really is magic; there may be hard times in the oyster business but there is no wounded dignity. Perhaps the tides wash it out. Or perhaps a propinquity to molluscs has healing effects that science has not even suspected. If so it would partly explain the qualities of the Postmistress too, since almost everyone she deals with is dealing in oysters.

For the unemployed intellect swatting at gargoyles there is no such salve. Madame bows before the blow and departs, having signed up for a guided tour of megaliths, one of several the director will direct during the tourist season. The Municipal Library of Vannes is by this time closed.

It is reached eventually, in the garret or servants' quarters of the Mairie, the imposing marble sweep of the first two flights of stairs giving way suddenly, through a small door, to a shabby little wooden third one leading up under the eaves. But this is proper. The true student wants no marble stairs and can do without a public toilet and a reading room; all he needs is books, and in this of all countries there must be plenty of those. There are some, on Prehistory; neither the volume by the local sea captain nor any other bearing on oysters is to be found. Impossible; here we are, in the very womb and capital of oyster culture, or at least of the culture of Ostrea edulis, and not a word about it appears either in the box of cards labeled Brittany or under Industries and Occupations. This last, along with a smattering of works on agriculture, electricity, general fishing and automobiles, shows instead an odd predilection: it lists forty-seven volumes on how to be a librarian. The librarian is apologetic; it appears that her predecessor for many years was an old man, that is he was old from the start and stayed that way, who didn't happen to have heard of oysters, or perhaps didn't like them. But no, way back in his early old age he must have; one volume of a general sort does finally turn up, published in 1890. The present incumbent kindly offers to borrow the sea captain's book from the Society of Many Learnings, to be read of course in that case at the table which constitutes the reading space as well as the space for unfolding maps, holding conferences, passing the time of day, etc. of the library. She is not sure the Director of the Society will be willing, and since the library has no telephone . . .

But it is only thirty miles and gas is only a dollar a gallon and the director does after all lend the book, which only involves three more trips, the library closing the first time for lunch and the next for a week's vacation, and once you get used to it the reading table is a pleasant place to be. It is sociable anyway, and an American student can feel at home; two out of three of the others at the table are looking at comics. One can imagine settling down to it, as an old oyster shell settles in the silt, and taking on through the years a slow accretion of prehistory and bibliographic science, safe from the crude practical world with its trade congresses and registration of births, deaths, marriages and traffic casualties in the marble grandeur downstairs. There are a few encyclopedias; one could move over to those after a while.

Only up there, nobody sings, and it is certainly not because of any demand for silence. A strange truth dawns, bringing together a long train of suspicions. Taking it all in all, the nation of Culture and the Great Exam honors the oyster a long way before the book. If it does so at times a little more flagrantly than the country of TV and the telephone, that is probably for two reasons. It takes a rich society to care about books, and in France the oyster is worth honoring.

BITTER tears were shed in many places over the slaughter of the oyster, but as usual, crying didn't help.

In the United States the process was marked by our customary efficiency and youthful enthusiasm; the relation of golden egg to hen was darkly hinted at in certain quarters, with no effect on practices in the business. The beds of Maine and New Hampshire disappeared entirely. In the gold-rush days oysters at $20 a plate were consumed in San Francisco at such a rate that the once enormous beds that provided them, in Willapa Harbor, then called Shoalwater Bay, in Washington, were also ruined beyond repair. ("A good specimen of the native oyster is a rarity there today." U. S. Fish and Wildlife Service publication, 1948) In spite of all warnings and evidence, the extermination went on at a crazily rising rate right through the century. For over a hundred years, beginning in 1820, the Maryland Legislature is said to have adopted more measures dealing with oysters than with all other topics put together. Result: nil. Chesapeake Bay, at one time the largest oyster-producing area in the world, went on being fished bald, with nothing done to replenish the stock. At the beginning of the 20th Century five hundred dredgers were working in the Maryland part of the bay; in 1947 there were enough oysters left to support sixteen. The beds of Georgia went the same way, losing quantity and quality together; the oysters got weak and tasteless as their numbers declined.

In 1943 another publication of the Fish and Wildlife Service estimated the take of oysters in the U. S. as "approximately half that of 50 years ago." Five years later it is the same story—"continuous decline" in South Carolina, Florida and Texas; "and the mortality of oysters in Louisiana waters threatens the very existence of the large oyster industry of the State. This critical shortage has developed as a result of many years of mismanagement of public oyster grounds and continuing destruction of bottoms by domestic and industrial pollution."

In one government report after another the same dismal vocabulary appears: destructive methods, lack of cultivation . . . Industrial wastes as from pulp mills, and harbor improvements in some cases, although damaging enough, get less blame than the chaotic system of regulations, different in every state and not thoroughly enforced in any, concerning both replanting and harvesting. For a long time many oystermen didn't even bother or were too ignorant to throw the shells back for cultch, or material the larvae must have to attach themselves to, and to keep the bottoms firm; they were sold to be burned for lime or used in roadbeds. Later efforts by people in the business and sometimes by state governments were often wasted, through lack of protection or of technical knowledge. In any case they were no match for the powers of blind acquisitiveness. The oyster population simply couldn't keep up with the haul, and where the beds stopped giving a quick profit, instead of being cleaned up and allowed to come back, they were abandoned. On the West Coast the problem was solved in Gordian fashion by importing seed from Japan, of a coarse species, Crassostrea gigas, fast-growing and used mostly in canning; it is very similar to, and may even be of the same species as the European Portuguese, although in our version it gets no such nurture as the best of those, so is scarcely comparable as food. This now represents 15 to 20 per cent of all U. S. oyster production, which continues to wane on the Atlantic and Gulf coasts.

Meanwhile American scientists have done some of the world's outstanding lab work on the oyster, as in marine biology in general. Among other items, the first successful attempt at large-scale artificial insemination of the oyster was carried out by two Americans, F. Wells and H. F. Prytherch, in the 1920's. It was backed by the federal government, tried out on a large scale, successfully, under the aegis of the N. Y. State Conservation Commission, and that was the end of that. Except in the laboratory nothing more was done about it for forty years. At present a single Long Island oyster company, one of the few remaining on the Sound, is rather secretively renewing the attempt.

On the destructive side, the European story was no prettier; legislation and human nature played the same blindman's buff. In the last century about as many oyster laws were passed in France, with about the same effect, as in Maryland. Voices of dire prophecy were heard in the land; there too the oyster was going, nearly gone.

In fact the banks had begun to be impoverished even in the 18th Century, especially on the Channel where great numbers of English boats came over to drag alongside the French ones. That is why our revolution was so popular in France; it kept the English sailors off the oyster grounds for several years, but the "pillage" began again in 1783. Not that there was anything to choose between the two as to the forms of pillage; the Norman and Breton fishermen just preferred, naturally, to kill their own hen, or cook their own goose. A distinguished naturalist of the time, the Abbé Dicquemare, sent by the government in 1786 to look into the situation, reported the banks in the whole gulf around the mouth of the Seine already diminished by half "in the last forty years." They had been exploited all year round, "in defiance of the laws of nature"; anything that could not be sold had been left to perish; women were selling oysters under legal size on the sly; and there were also the various frauds and abuses of *la pêche à pied*, resulting in total destruction of mothers, offspring and all. "The real causes of the deficit," writes the Abbé, "are the maneuvers of cupidity and the insufficiency of laws."

So of course some more laws were passed, involving limits on the fishing season, strict supervision and sorting of the catch, payment to fishermen for the work of returning small oysters to the beds, etc., in spite of which the banks off Cancale in the Bay of Mont-Saint-Michel, formerly fished to the tune of over 100 million oysters a year, were giving only one million in the years around 1800. They were later given a rest and reconstituted, and like many others rescued in the same way, were reduced again about as soon as they were reopened.

One protective measure, applied still earlier and not on the Channel, had a curious result. In 1750, or less than seventy years after the banks had been officially called inexhaustible, they were in such a bad way in the Basin of Arcachon that the Admiralty forbade all oystering there for three years, and for once even the fishermen, with a foresight unique in the general picture, did everything to enforce the measure instead of trying to get around it. It was like the magic porridge that grows to a torrent in the fairy tale. Before the time was up the oysters had multiplied until they were a menace both to navigation and to public health; they stacked up in reefs in the channel and so far inshore that many died from lack of water between the big tides, making the air putrid with their decay. A contemporary account states that in spite of this brilliant result, the Admiralty was under no illusion that anything would be left of the bounty before long, without strict and constant control. In 1755 the Parliament of Brittany forbade all dragging on the Tréguier bank except at Lent, and all exportation of its oysters by ship. As a result of lobbying, the prohibition was lifted after only three years, and within six the bank had been ruined again; new six-year prohibition, this time by the King, under which the banks revive, to be stripped again five years later; new regulations, requiring all drags to be left at the Hôtel de Ville during the closed season; another period of health, followed by another extermination.

That was the story more or less everywhere, and although by the early 19th Century these early hit-or-miss rulings had given way to a more rational and general control, or attempt at it, by and large the banks went right on vanishing, through that century and ours. A hundred and fifty years after the Abbé Dicquemare's report, exactly the same complaints are being made, although by that time dragging was being permitted in many places only a few hours or even one hour in the year. The draggers are not returning young oysters to the bottoms; on the public grounds, as distinguished from parks, nobody, including those whose whole livelihood is at stake, is interested in helping to clean the beds, or as in one glaring incident in the 1930's, to rid them of starfish.

A report of 1858 to the Emperor, His Majesty Napoleon III, on the oyster beds of France, written by a man who is going to appear again in this story, sounds like a knell: "Weakened from Cancale to Granville; extinguished at La Rochelle, Oléron, Rochefort, Marennes . . ." This is signed, like all the man's works, "Coste, Membre de l'Institut," without any initial. In another report he goes into a detail of the process.

"In setting the opening of the campaign in early September, the ruling has had a certain wisdom, since parturition is mostly finished at that time, and there is not much risk of pulling from the water mothers having their progeny still within them. But this progeny, which before expulsion formed inside each 'milky' oyster an innumerable family, spreads around afterwards on the outside of the valves, becomes encrusted on it and forms a new population on the surface of the old. Now if, just as this replenishment occurs, the bank is opened to exploitation, the damage will be almost as great as during gestation, for along with the adults it removes the coming generation that they carry, that is everything that has not deserted its source. The drag will therefore devastate these fields in full germination, like a rake passed through the branches of a tree in flower. It is one of the causes, and not among the less serious, for the impoverishment of the coast." He adds that in any case, "six weeks of daily dragging would be enough to denude the whole coast of France."

Coste never attacked the use of the drag as such. Some other people called it the oyster guillotine and every other kind of bad name, for its indiscriminate ripping up of the beds and sometimes of the beautiful and necessary fields of vegetable stuff on the floor of the sea, which keep the mud from piling up over the beds in certain places. It consists, as it did then, of a convex iron blade of a length usually subject to law, sometimes five or six feet, serving as a scraper, and behind that a pouch formerly of net or thong and now most often of metal mesh, into which as many as 1,000 or 1,200 oysters may tumble at each dip, along with a swatch of marine garden, shells, mud, gravel and anything else in the way. Where beds are "cleaned" in advance, as the Long Island ones are now, this of course doesn't apply. The drag is pulled along behind the boat by means of chains, and considering the great depth of water it is often used in, and the fact that some French draggers even now have no winches, represents a rather extraordinary feat of manpower. There were arguments for it, if it were used wisely; the beds were often in need of the cleaning it could give, and it was at least as likely to keep the mud down as bring it up through weed-pulling; for all the griefs and diatribes it brought on, there was probably nothing inherently evil in it, or nothing to outweigh the gain. It is just another of the useful human tools, a very modest one in this case, that the devil may take a shine to.

The region we are concerned with went with the times, like the rest, only somewhat later, by which we see that association with witches, menhirs and even saints is really of no practical help, and at least in some respects, a Breton will behave just like a Norman or a Gascon or an American if he gets a chance. Around the corner from Locmariaquer, at La Trinité where they are so particular about how you eat your ice cream, 126 oyster boats were working in 1881, and 25 in 1890. In the River of Auray, a dragger in 1878 was taking in an average of 742 oysters an hour; thirteen years later the hourly haul was down to 69. In 1911 some of the beds closest to Locmariaquer were found to have vanished or nearly so, and the great offshore bank of Quiberon was also on the way out, and it seems that a ruined oyster bed is a most dismal scene. Starfish and oyster drills come swarming in, mussels may multiply on the empty shells or smother the last of the living oysters, the ocean floor once a pasture and kept firm by new generations of shells is covered by shifting sand or mud, in which black sea worms burrow for a dying oyster or corpse of one; the last sad progeny of the race is eaten instantly if it is not washed far away, until finally, where millions used to grow, not a single young one is to be found alive.

This was the horrid spectacle that by the year 1900 lay at point after point along the French coast, not to speak of our own, beneath the keels of pleasure boats and the lyrical drift of bright-colored sails in the sunset and the heels of travelers, massed at their ships' rails for the approach to La Havre, Cherbourg, Dieppe, almost any port from Calais to Bordeaux. For almost all had had their oyster beds, if not in the port itself, at least nearby.

Not that they weren't still available. Some 564 million are said to have been sold in France in the season 1880–81; and here is what you had to choose from at Les Halles in Paris in 1890, with prices in francs per dozen and per hundred:

	Doz.	100
Portuguese green or white	.40–.90	3.25–6
Arcachon, small	.30–.80	3.25–6
" middle-size	1.	7
" big	1.25	8
Armoricaines, small	1.60	10
" middle-size	2	12
" big	2.50	16

	Doz.	100
Belon	2.50	18
Marennes, white 2nd	1.25	8
" " 1st	1.60	9
" " extra	2.00	13
Sables d'Olonnes	1.60	10
Marennes green, ordinary	2	14
" " extra	2.25	16
" " big	2.75	18
Cancale ordinary	2	13
" extra	2.25	16
Pieds de cheval	3	22
Courseulles	2.25	15
Ostend	2.50	18
" extra	3	22
" Victoria	2.50	20
Zeeland	3	22
Royal Whitstable	3.50	25
" " extra	4.50	32
" Colchester	3.50	27
Burnham	3	22
Second Native	2.75	20

That's nothing compared to cheeses, of which there are either 289 or 300-plus kinds made in France, depending on your authority. Still it must have been rather bewildering. All of them, with big differences in fattening and refinement, according to degree and place of élevage, were of the two species *plate* and Portuguese, the latter having appeared in France only twenty-five years earlier. There would have been, and still is, a further confusion from the use of the term *huître blanche*, a term occasionally used for the *plate* but which in general merely means an oyster, of either kind, that is "white" because it has not been made to turn green, as at Marennes. The careful reader will observe that many on the list were imports from Belgium and England, not that they necessarily started out there. Another point is that many places known as great oyster centers, notably at that time Courseulles, had and have now no beds or other possibility of producing oysters of their own; they are in the business of raising and improving, and have to buy their supplies elsewhere, as our West Coast oystermen do from Japan. With all the natural sources of supply in more or less the same distress, there was therefore a great deal of mutual scrounging around for young oysters to keep these establishments going. In fact if that were all there was to the story they would not have kept going and it would have been very poor pickings at Les Halles. As witness a complete rundown of the banks published in the 1930's—not by a scholar, not by the government. It is by a passionate old oysterman of the Marennes region, named Paul Hervé, in a book less stirring by its facts than its vivid ramshackle eloquence, and the fact of such a man, not illiterate but scarcely handy with the written word, having taken the heroic pains to write it. And what ghosts and echoes rise from every one of these names, of old sieges and naval battles and landings in the last war, of Henry Adams stalking his other ghosts around Mont-Saint-Michel and worrying about everything but what the boats were up to under his nose; of Proust tormented by the fluttering skirts of girls on bicycles at Cabourg, of impressionist painters and left-bank writers in the 1920's still able to find the perfect little place for the summer; of King Arthur, and the first Celtic saints, and our mothers and grandmothers deliciously bowing to their foreign acquaintances in the hotel dining room by the sea. Their faces wobble and liquefy; we are looking up at them through water, in this history.

The so-called Great Bank, off Calais, "completely vanished as of fifty years ago"; Boulogne—several parks abolished, one remains, where visitors can déguster the product; Dieppe—"the numerous and very ancient parks, whose oysters were renowned, are abandoned today"; the oysters of Étretat near Le Havre, favored by Marie Antoinette, "are today only Portuguese"; off Le Havre, the Great Basin and the Little Basin, once a paradise for oysters by the tremendous affluence of fresh water there, except at one point now have only an occasional stray; reservoirs abandoned at Harfleur—"to sum up, this part of the channel no longer embraces anything but ruined banks"; the section around Caen, another paradise by virtue of two river-mouths and configurations of chalk rocks—natural banks "annihilated," most parks given up, one bank extending ten miles offshore ravaged by English in spite of a convention of 1839, another taken over by mussels, another disappeared in 1875; Courseulles, renowned in oyster annals, steady decline since 1890; Saint-Vaast, neighboring Cherbourg, another hoary name, saved to some extent by "the culture of the Portuguese," which however does not reproduce there; Cherbourg—banks "no longer exist"; Barfleur—"dragged white," the bank rapidly disappeared; Granville—parks closed because of contamination; main bank of Saint-Malo, once furnished 20 million a year, "none now"; ports of Saint-Malo and Saint-Servan—"consommation interdite"; mouth of the Rance, "once extremely fertile, now only one small deposit"; Saint-Brieuc, once very rich, "nothing left"; Paimpol—"ruined"; Tréguier—reduced by half since 1870, parts near city closed for contamination, no naissain . . .

Et ainsi de suite, all the way on around Brittany and down to where the Loire along with its mud seems to pour reflections of châteaux and deliquescent portraits of kings' mistresses eternally into the sea, and on past the Gironde and Arcachon and over to the Mediterranean, which doesn't matter so much.

But something else was happening at the same time, making the main difference between this picture and the American one. It began in the 1840's and its great single hero was Coste (1807–73), embryologist and ichthyologist employed by the strange government of the Second Empire and known as the father of modern oyster culture. His initial seems to have been V.

FOUR

ACTUALLY CERTAIN foresighted oystermen had been working on the problem for several years before Coste came into the picture. It was then perfectly clear, except to most people, that natural propagation alone was nowhere near up to the demand, and probably wouldn't have been even with the best of regulations and no infringements or abuses, but that would have been rhetoric; no government could afford a marine police force on such a scale, and people with a living to make were not going to turn virtuous to help save a natural resource, or help somebody else make a pile at their expense. What had to be discovered was a technique of breeding, meaning, in this context, collecting larvae and raising seed.

It is astonishing how hard this was, and how new, considering all that was known about oyster cultivation in general. That part of the business, highly developed in France long since, was almost as old as history.

The ancient world, while knowing nothing about the reproduction of molluscs, had known a good deal about the necessary conditions for it. Aristotle, who also did some speculating on the sex angle, tells the history of oyster-raising "as now practiced," from the discovery that oysters had attached themselves to the broken pots and other bric-a-brac thrown overboard by the fleet at Rhodes, to this, touching on a key question of élevage: "Some fishermen of the island of Chios, having taken some oysters at Pyrrha, on the island of Lesbos, and having taken them to another place in the sea nearby, where the waters formed a current, the oysters fattened greatly but did not reproduce at all, although they remained there a long time." Since boys will be boys and the ancient world was strewn with broken pots they presumably went on being used for the same purpose. Pots or not, toward the end of the 2nd Century B.C. an enterprising Roman named Sergius Orata latched onto the idea of cultivating oysters artificially and soon was making a fortune at it in the vicinity of Baia, near Naples.

An interesting if all too familiar type, this Orata—tycoon of the luxury business; there must be a bust of him somewhere. The following is from Coste.

"At the bottom of the Gulf of Baia, between the shore and the ruins of the city of Cuma, there can still be seen, inland, the remains of two ancient lakes, the Lucrinus and the Avernus, formerly joined by a narrow canal, of which one, the Lucrinus, had access to the sea through an opening in a dyke traversed by the Herculean Road; peaceful basins, which an upheaval of this volcanic soil has almost totally filled up, and where, as the poets said, the sea seemed to come for rest. A crown of hills bristled with wild woods casting their shadow on the waters, had made them an inaccessible retreat, dedicated by superstition to the gods of the underworld, and to which Virgil led Aeneas. But around the 7th Century [from the legendary founding of Rome] when Agrippa had stripped them of this gigantic vegetation, and the subterranean road (grotto of the Sibyl) had been dug leading from Lake Avernus to the town of Cuma, the deshrouded myth gave way before the works of civilization. A forest of magnificent villas, built and decorated with the loot of the world, replaced those somber groves. All Rome descended on this pleasure spot, drawn by its balmy skies and azure sea. The hot springs, sulphurous, aluminous, salt, nitreous, overflowing from the mountaintops, became the pretext for this displacement of patricians, driven by boredom from their homes.

"Every resource of industry was tapped to surround them with the pleasures their softness craved, and among those who threw themselves into this enterprise, Sergius Orata, rich, elegant, attractive in manner, and of great influence, hit on the notion of building up oyster parks and making this mollusc famous. He had his oysters brought from Brindisi, and persuaded everyone that those he raised in Lucrinus acquired a better taste than those of Avernus or even the most famous regions.

"He succeeded in selling the idea so rapidly, that in order to keep up with the consumption, he ended by filling almost the whole circumference of Lake Lucrinus with the constructions he kept them in; thus taking over the public domain with so little inhibition that suit had to be brought against him, to take back what he had usurped. At the time when he suffered this mishap, and to show how far he had perfected the business, it was said of him, in an allusion to the penthouse baths he had also invented, that if he were prevented from raising oysters in Lake Lucrinus *he would find a way to grow them on the roofs.*"

There is a slip in chronology here. According to Pliny, Orata functioned in the time of the orator Crassus—not Crassus the Rich, of the triumvirate, but the earlier one, before the Marsic War of 90–89 B.C. when the Italian peoples fought for the rights of Roman citizenship. Agrippa's great engineering work that transformed the landscape and waterways in that neighborhood was undertaken to make a harbor for Octavius' fleet before the Battle of Actium, or two generations later. The place, only a short trip from Pompeii, went on developing at a great rate afterwards but before Agrippa's time was already becoming the kind of resort Coste describes, and in fact the name Lucrinus, meaning literally, with beautiful simplicity, Money—from *lucrum*—is said to have come about from the fortunes made there in the oyster business. Nobody, except perhaps Virgil, seems to have gagged on either the name or the oysters, or the lake's being indeed linked by a narrow canal with the entrance to the underworld. Avernus, the other lake, was said to be two hundred feet deep and so miasmic it killed birds flying over it; all was tenebrous and fateful under those azure skies; perhaps it was a taste of pomegranate that gave the oysters their distinction. A third lake nearby, inheritor of Orata's business many centuries later, the *palus Acherusia*, was also linked to the regions of the dead, like its eponyms in Egypt and Epirus. And the myth was deshrouded only for a certain crust of society. Everything supported it geologically in that black-soiled, bubbling, unstable area; secret sacrifices were still being made on the shores of Avernus long after Constantine.

We remember Aeneas' approach, of course not *led* by Virgil as Coste puts it, but pushed, that is dreamed up by him, or more accurately, remembered, for the only real vice of any society is lack of memory and the only real function of a poet is to supply it. The guide is the theatrical and athletic old lady the Sibyl of Cumae, who does everything to make the journey as scarey as possible, quite properly. The poignant note is the golden bough that Aeneas must find first in the dark forest in order to be admitted to Hades, and his joy and relief in being helped to find it by the two white doves sent by Venus, when in fact at the time when Virgil was writing, the way things were around there, a golden bough would have been no more than you would expect to see in anybody's back yard or atrium. The earth rumbles and quakes and dogs howl in the darkness while the Sibyl slaughters her black bulls at the entrance to the terrible cave and with a wild shriek plunges down the road, which is not paved with good intentions as in our mythology but is just as unpleasant in other ways, and so on to the junction of the two great rivers of the dead, the Cocytus named for wailing, and the Acheron, where the stingy boatman, originally from Egypt, picks and chooses among the mobs of shades waiting to be ferried to rest.

That spot would be directly, give or take a few miles, beneath the oyster parks, and not far from the house Pliny tells about, where the owner became so enamored of one of his lampreys he wept when it died; at the same villa a lady named Antonia adorned another favorite lamprey with earrings and its reputation made some people extremely anxious to visit the place. Cicero had a country house there too and no doubt some other serious people, but it can't have been a very good place to work. On the whole it is pretty much as if the gates of hell were situated in Westchester County or Beverly Hills.

This was the setting, perhaps not of the western world's first deliberate cultivation of oysters, as so often claimed, for the same thing seems to have been done earlier at Taranto, but at least of its first really flashy oyster business. We don't know how Orata ended up; he was clearly one of those men in whom the passion for making money is equaled only by the knack for it, and it doesn't sound as if a mere lawsuit could have hurt him much. Lake Money, which symbolically ought to have erupted over his business practices and other wickedness the way Vesuvius did across the bay, didn't get around to it for another fifteen hundred years. So much for divine retribution.

The other important question is exactly what Orata did with his oysters, whether or not he brought them from Brindisi. Coste may have been a little hasty in his reading there, as on Agrippa. According to Pliny that was a later development, as was the importation of oysters from Britain. There evidently were natural oyster beds in the lake, which was just enough open to the sea to allow for it, and his innovation was either in taking the young oysters from their clumps and subjecting them to the process of élevage, or in that plus breeding, that is collection of seed. Whatever he did it made a dazzling impression on his time. He soon had a bevy of imitators, and it appears that the far more elaborate *ostrearia* of the early Empire did involve the use of collectors. Two glass funeral jugs of Nero's time, one of which was in a museum in Rome in Coste's day and may still be if it hasn't been misplaced or broken by the cleaning woman, show oyster basins connected with some quite fancy buildings and communicating with the sea through a series of arcades, also what look like an arrangement of pickets in the water, which Coste and other people after him have taken to be collectors.

Incidentally Pliny, more an egghead than a nature-lover, had an interesting view of oyster procreation. Following Aristotle, he says it comes about through spontaneous combustion in the mud.

OSTREOLOGISTS, like most other ologists, tend to hurry over the next fifteen hundred years or so, the way you hurry down a lonesome road at nightfall. Some form of artificial cultivation went on all the time, here and there. We hear of it in Denmark in the

10th Century, and in Schleswig around the same time; a 16th Century writer speaks of its having been a practice from time immemorial on the shores of the Bosphorus. Rather poor pickings. But one item jumps out at you. It seems that in England in 1375, in the reign of Edward III, a law was passed forbidding the transportation of seed oysters except in May. Of course that is only startling to us mortals; it wouldn't be to a medievalist.

Leaving out the experts, nothing is as private as one's sense of the Middle Ages. Nowadays you might even say it is the only privacy left, so it would be a shame to give it away. At the moment all we care to divulge is a feeling of dismal darkness over the whole period, which no amount of architecture or other fact or fancy, or even the blazing, abnormal sunshine of *The Song of Roland*, has ever dispelled. This only proves the pernicious power of phrases, "Dark Ages" in this case, but there it is—the living embodiment of the death rate, with the Black Prince twelve feet tall striding through a dark field of brambles called the Hundred Years' War, and a few lucky knights off on stag parties to the Holy Land, accompanied by the usual plague, sack and rape but at least able to see their hands in front of their faces. Of course Dark and Middle are supposed to be two separate pieces of time but they weren't in our education, and just where the fracture comes doesn't seem to be too clear anyway.

Enough; shame forbids; besides, we're waiting for an offer. We'd bet a lot, though, that Henry Adams started out with some such image, and that was what drove him to all his spiritual hassles over Mont-Saint-Michel and Chartres.

Well, Mont-Saint-Michel used to look out over some of the most prolific oyster beds anywhere, so we're at least in a respectable neighborhood, intellectually, and it's astonishing how that edict does brighten things up. You'd think the clerk would have put an exclamation mark after it, but no, he didn't see any reason for emotion. Perhaps he should have. The Black Prince, who was of ordinary size for the time, that is rather short by our standards, and not black in the face, it was just the color of his armor, was going to die the next year, and his father Edward the year after, leaving a rather bad situation for a number of years. This was not just *in* Edward's tremendous fifty-year reign but two years before the end of it, and right in the thick of the briar patch, which as every schoolboy used to know went on well over a hundred years and in 1375 still had a long time to go.

It was a year of peculiar suspense on both sides of the Channel; in France too a strong monarch, Charles V, island of sense in the anarchic anguish of the 14th Century, was to die within a few years. The English, having only recently finished conquering almost all the Atlantic oyster beds of France, had already been beaten off quite a few of them. They were getting tired, the House of Commons was grumbling over voting for war taxes. Harassed on land by the Breton Du Guesclin, Charles's great military man, and on sea by the Spaniards and corsairs working for the French, the English were being pushed back to two last enclaves, Calais, which they kept for another two hundred years, and the Bordeaux region—both rich in oysters.

Two of the big oyster centers of Brittany had figured in the deadly wrangle a little earlier, in a more complicated way. The long side-war between the rival dukes of Brittany, Blois and Montfort, backed by the French and the English respectively, came to an end with the Battle of Auray in 1364, a victory for England's team; the Blois man was killed and Du Guesclin taken prisoner but he soon bought his way free. That was in September. That same year Du Guesclin, hero of Brittany for the time being, had been operating in his home district around Tréguier on the north coast, where English castles, we are told, were perched like vultures' nests on every crag and mountain. A troubadour of the time has the local townspeople appealing to him.

> Ah! Lord Bertrand, God's blessing on you!
> We have sore need of you, in my opinion;
> For there are castles full of English
> Who every night come right into our courtyards.
> They make off with our cows, sheep, lambs;
> From Castle Pestivien the worst comes to us.

Actually, the "troubadour" in this case is a 19th Century Breton aristocrat, poet and *homme de lettres*, named Hersart de la Villemarqué, figure of extraordinary controversy; of which more later. For the moment it will be more edifying to duck the brickbats and go along with him. The following, also from the

French in his great *Barzaz-Breiz*, or supposedly popular songs of Brittany, for which he also gave Breton versions and music, tells the end of the same castle, Pestivien. It is described as being in "la terre des Anglais," with deep water all around and a tower at each corner.

> In the main court, a well full of bones, and the heap grows higher and higher every night.
>
> On the grill over the well the crows swoop, and they go down to the bottom of the well, looking for fodder, croaking joyously.
>
> The drawbridge of the castle falls easily, but even more easily is raised; whoever goes in there will not come out again.

A young equerry wanders in unawares, and having avoided being murdered through the help of a servant girl, rushes to town to tell Du Guesclin, who cries out: "By the saints of Brittany, as long as an Englishman is alive, there will be neither peace nor law!" and razes the castle to the ground. This had some basis in fact and so did another story of the kind, subject of another ballad, in which the English villain tries to rape a peasant girl. She prefers suicide to dishonor, and as she happens to be Du Guesclin's goddaughter, the future *connétable* is quickly on the scene.

> Rogerson has been killed, the castle of Trégoff destroyed;
>
> Destroyed is the castle of the oppressor; good lesson for the English!
>
> For the English good lesson! good news for Bretons!

Fifteen years later, the tune has changed. Jean de Montfort, not long after his splendid victory at Auray, had had to flee to England, he and his English friends. But now we are in 1379, or La Villemarqué's version of it:

> A swan, a swan from overseas, on the top of the old tower of the castle of Armor!
>
> Ding, ding, dong! to combat! to combat! Oh ding! ding! dong! I am going into combat! . . .

Lord John has come back, come to defend his country,
Defend us against the French, who trample on Bretons.
A cry of joy rises, making the shore tremble; . . .
The bells sing joyously in every town, a hundred leagues around.
Summer returns, the sun shines; Lord John has come back!
Lord John is a good friend; his foot is as quick as his eye.
He has sucked the milk of a Breton, milk healthier than old wine . . .
The hay is ripe: who will scythe? The wheat is ripe: who will harvest?
The hay, the wheat, who will take them away? The king claims it will be he;
He is coming to scythe in Brittany, with a silver scythe;
He is coming to scythe our meadows with a silver scythe, and harvest our fields with a sickle of gold . . .
The wolves of lower Brittany grind their teeth, hearing the call to war;
Hearing the joyous cries, they howl: at the smell of the enemy, they howl with joy.
On the roads, blood will soon be seen flowing like water,
So that the feathers of ducks and white geese swimming past will be red as embers.
There will be more broken lances than branches on the ground after a storm;
And more skulls of dead men than in all the ossuaries of the country.
Where the French fall, they will go on lying till judgment day;
Till the day when they will be judged and punished together with the Traitor who is leading the attack.

The traitor: our old friend, flower of chivalry and scourge of the enemy, Du Guesclin. They said Charles V had blinded him with that title of *connétable* and other honors; he had forgotten where he came from. Kicking out the English had been all very well, as long as nobody else was at the gates. Now Bretons everywhere, even his own family, were turning their backs on him. His own soldiers left him to go over to the army of Brittany, and although he remains the great hero of French history, some said he died of grief and repentance for having led a French army against his own people. Meanwhile, on the occasion celebrated by the ballad, peasants, bourgeois and nobles fell over each other into the sea to greet the return of the previous traitor, Montfort, whom they had called back. His worst enemies of the decade before, even the widow of the rival he had killed, Charles de Blois, knelt before him as the liberator of Brittany. According to a poet of the time, he wept as he helped them to their feet, and without further ado set off against the French, followed henceforth by men *born in Brittany.* At any rate that is the picture we get from the notes to the *Barzaz-Breiz.*

Between these two main events, the Battle of Auray and the return of Montfort, the English government took the time to pass the aforementioned law about the transportation of seed oysters. Just what they did with the seed they had transported is not clear; cultivation in parks is said to have started in England around 1700, not before. Nor is anything said about oysters being brought over from France but very likely they were, between skirmishes. At least the law indicates that the business was on a scale worth bothering about, and that for all its deadly upheavals Britain had kept up the Roman practice of raising oysters away from their original beds.

Chaucer, by the way, was working in the government wool office down the street at the time, in a rather lordly capacity to be sure, and writing about the "grisly feendly rokkes blake" of "Armorik"—

> That semen rather a foul confusioun
> Of werk than any fair creacioun . . .

and of Breton songs a long way from the spirit of Trégoff or Pestivien or the Battle of Auray:

> Thise olde, gentil Britons, in hir dayes,
> Of diverse aventures maden layes,
> Rymeyed in hir firste Briton tonge;
> Which layes with hir instruments they songe,
> Or elles redden hem for hir plesaunce . . .

How bright and peaceful. The husband in the story goes off to England "to seke in armes worshipe and honour" merely in the normal course of chivalry, with no international complications at all; and the attempt at villainy in his absence ends in an inner triumph of the most tender generosity, as if there were no horror in the world but those black rocks.

IN France the park system was highly developed in the 18th Century. In Normandy it had to be, to keep the court supplied; there were two hundred parks in Courseulles alone. In Marennes the process of "greening" in parks was established well before that, in spite of its being in a virulently Protestant region and therefore with limited sales possibilities; too far away to provide the Paris market, it had no sooner got through with the English unpleasantness than it was caught up in the religious wars, which didn't increase its outlets. Louis XIV, on a snooping expedition in the neighborhood, was offered some of these carefully nurtured oysters, later prized above all others in France but only a local delicacy at the time. Mme. de Maintenon, alarmed by their color as Americans still are, thought the king was being poisoned and had them taken away. However, it was explained that this was a specialty of the *claires*, or oyster basins, of Marennes, a natural product and in no way harmful, so the royal party was not deprived after all. The gossip around town was that Mme. de M. hadn't really thought they were poisonous anyway but was just venting some of her passionate hatred of Calvinists. A large proportion of those oysters even then were not native to Marennes, but were brought at some stage in their youth from Arcachon and other places.

To sell them, until well on in the 19th Century when the roads began to be improved and the railroads were extended to that part of the coast, the wives of oystermen would spend the whole winter in Bordeaux, from early September through March, usually without being able to see their husbands and children at all. Several times a week the husband sent a shipment downriver and the wife would be at the dock to meet it. Each one had her own "gate," a stall with luck but more often a spot in the street, with or without protection from the weather; if it was at the door of a restaurant they paid something for the concession and ran errands for the proprietor. It also improved the tone of the restaurant to have its own *écailleuse* at the door. One result of this economic necessity was that the women quickly became citified and stopped wearing head shawls, so that the tone of the home town was never the same again. It served to point up the importance of marriage too; a bachelor oysterman was almost a contradiction in terms.

The women of Arcachon did the same thing but were near enough to Bordeaux so that it was not on any such heroic scale; they could go back and forth every few days. There as in some other places, the oyster parks had developed more or less accidentally. At Marennes it was most often in old salt basins, the *marais salants*, given over to it when oysters became more profitable than salt. It was a way of keeping whatever couldn't be sold the same day, and only gradually developed into a deliberate business of raising and improving the oyster. At first people staked out parks wherever they cared to and could; places closer to shore were easier to tend and to get to, but the oysters were also more vulnerable to extremes of temperature, so in spite of the extra labor many oystermen pushed out to areas uncovered only at the great tides. Government concessions at last became necessary but for a long time were free and hereditary; there was no fee for use of public waters until the middle of the 19th Century.

IN short, while the right hand destroyed, the left cultivated. The tricks of that part of the trade were well known. It seems that this is not common knowledge even among some oyster fanciers so it is worth emphasizing, to wit: people in many parts of the world had been improving oysters by raising them in other than their native waters for at least two thousand years.

Raising seed, however, was another matter. It was being done in China, by means of stones and woven bamboo, and in Japan, and at certain points on the coast of Mexico, where Indians used branches of trees as collectors, and in Italy, at Taranto and in Sergius Orata's territory. In the United States and most of Europe nobody knew anything about it, and nobody anywhere knew *how* the oyster reproduces itself, any more than they had in Pliny's time.

Or no, more than that was known, not widely though. Coste and a few other scientists were getting warm, as you might say, about the sex life of the oyster. Science was young and passionate in those days, so short a time ago; the great truths, the Truth itself, flickered grail-like behind every leaf and shell and over the grind of the lab, where the coming race of mere technicians hadn't yet cast its pall. This is where our time sense goes most crazy; it's an old inner tube with a couple of dogs quarreling over it; if it's a yardstick it will wrap itself around the world and be insufficient to measure the common caterpillar. Geological time, with margins of error of mere millions of years, is more comfortable to move in than the last hundred and twenty-odd, if you start imagining one by one all the hunches, experiments, discouragements, hours of ghastly boredom and everything else on the part of a number of people, that went into finding out anything about anything, such as the births of mountains or how the oyster procreates.

And it goes on. Pop brontosaurus into extinction, whisk the incipient whale back into the sea, stick wings on the snake like a mustache on the billboard lady, rather than think of the little government office in Auray, which like four others in France and numberless others around the world conducts investigations all the time—on the effects of temperatures, salinity, currents, rivals, diseases, etc. on the oyster, on the same assumption that is made in human biology, that the animal serves some purpose and even if it doesn't it is something to find out about. And we are committed to finding out. That seems to be the new fact of life. We have lost the passion, by and large people are as jaded by the progress of knowledge as by other excesses, but we are hopelessly committed.

So even the consoling oyster, and quite aside from its propensity to disappear when poisoned or otherwise mistreated, is not really a dependable creature; something surprising will very likely be learned about it tomorrow. Still, a few facts have been established in this long, long century and a quarter.

To begin with, like many bivalves it is a hermaphrodite, as everybody knows but didn't in the 1840's. Coste was among the first to say so, although he was wrong in thinking it self-fertilizing, i.e. that the male and female parts worked at the same time in the same individual, a set-up not unknown among molluscs but rather rare and apt to be only an occasional event. Studies by other scientists followed and in the early 1880's, with the work of the Dutch zoologist Hoek, following an earlier French lead, the cyclical, or alternating, nature of the oyster's dual sex life was determined. Nevertheless, in case all this should sound like smooth sailing, in a report of the U. S. Commissioner of Fisheries written in 1919, we read: "The sexes are separate. Some oysters are male, the reproductive organs developing spermatozoa or milt; other oysters are female and produce ova or eggs. While it has at times been stated that the sex might change from year to year, an oyster being perhaps male one year and female the next, or vice versa [sic], there is no evidence on which to base this belief, except some inconclusive researches made nearly fifty years ago and not borne out by subsequent investigations." The author of this document also states that "fertilization of the eggs occurs in the water." Within a few years of that writing Hoek's findings were borne out by several investigations (R. Spärck in Denmark, 1925; J. H. Orton, 1927, 1933, and H. A. Cole, 1942, in England); it was proved that the oyster changes its sex not only from year to year (or vice versa) but often within the same year. As for fertilization of the eggs in the water . . . But you already have that, on pages 23-4. (Crassostrea yes, Ostrea no). Mr. Philpots had been referring to all this as a matter of general knowledge in 1890.

We don't know what happened to this vice-versa fellow, whether he was fired or promoted or committed suicide, and don't want to know. To do him what justice is due, he probably had in mind the fact of protandric hermaphroditism, meaning that the first sexual activity is as a male.

Secondly, the oyster is prolific to the point of indecency, with great variations depending on age, size, season, exact place and species, the *plate* being low in the scale. It is now known that a single female, for the season, virginica may discharge from 15 million to 114 million eggs in one spawning, whereas a million is average for a healthy edulis four years old, the age of maximum reproduction. Incidentally, the phrase "mother oysters"—*huîtres mères*—that still turns up in much literature on the subject is of course a leftover from the old days, except as applied in a specific moment. You can't speak of putting mothers in a certain location when they may turn into fathers on the way. It should be "parent" oysters, or just adults.

Since our heroine, when not being a hero, keeps its eggs through incubation, many are "milky" in the laying season, which also varies a good deal but is generally at its height in the Morbihan in June and early July. After fertilization the larvae stay put for a week or ten days, the milkiness turning different shades of grey at this stage, and are then shoved out into the world where most are promptly killed in one way or another. Those that survive enjoy for a week or two the only freedom of action they will ever know. This is known as their free-swimming or pelagic period; they are carried here and there by currents, but grow a little hairlike rudder that gives them some say in the matter, and also, at the end of this vulnerable but perhaps happy time, a tiny foot with which to attach themselves to something, which they must do when the moment comes or die. Once "fixed," as the process is called, the oyster loses its foot and swimming apparatus and will never move again under its own power; an awesome requirement, but then all nature fixes in some fashion, even if only in being existentially "engagé," and by and large there seems to be about the same proportion of will to chance in human fixations as in those of the sessile mollusc. Freedom, as they say, is relative. What is impressive in the oyster is that it learns this grim lesson so young. It is at this point visible only with a magnifying glass, yet equipped with a secretion such as many humans take half a lifetime to acquire if they ever do, whereby to attach itself *in perpetuo* to the object of its fancy. That is, barring outside intervention.

It is not particularly choosy, will fix on a rusty nail or a shell or an old bit of wood, but is not altogether indiscriminate either, having apparently a slight preference for clean surfaces and especially for lime. In natural oyster beds the larvae have little choice, fixing mostly on the shells of older or dead oysters, which makes for various inconveniences both to themselves and to the oysterman. Many oysters become misshapen or under-nourished from overcrowding, and the older ones may be smothered by the new crop. The clusters when dragged in, being made up of oysters of all ages along with the shells, require far greater work of sorting and cleaning than those from artificial beds, and many of the young and fragile are broken in being pried away from the rest.

Hence, in the case of Ostrea edulis, the tiles. It took about twenty years from the 1840's to the 1860's to work out the system, which has been in use ever since, and the pride of the Morbihan in the matter is justified, although it has to be shared as to origin with Arcachon. It was there that the first government concession for the purpose of seed production, as distinct from cultivation, was given in 1849. That attempt and several later ones, all private, were failures; various pleas for help from the state went ignored. But at last, in 1853, the government sent Coste to study the methods in use, not in Orata's lake but a few miles away in the old Infernal Pond, *palus Acherusia*, under its modern or Christian name Lake Fusaro.

Lake Money had been blown up by a volcanic eruption in 1538 and partly replaced by a mountain called the Montenuovo. History gets a bit hazy here but oystering seems to have limped along somehow in the neighborhood until the 18th Century, when Ferdinand Bourbon of Naples got it off to a fresh start in Fusaro. The peculiar thing is that this fascinating industry, in one of the most be-touristed spots of Europe, should have been so obscure, if not totally unknown, outside the neighborhood in Coste's time. It came as a huge surprise; nobody had mentioned it, nobody cared. The beautifully educated travelers of the 18th and 19th Centuries, who after their fabulous days of horse travel, sight-seeing and parties at embassies and princely palaces, still found time to compose volumes of letters and diaries every night, went to that region for Pompeii and the bay and whatever was being done on the mandolin before "O Sole Mio" (1898). Another project for scholars: find one in the lot who ever referred to the parks of Fusaro. Yet they would take the most exhausting trips to see the lower classes making lace, blowing glass, etc. They knew their Virgil, and probably they *ate* the oysters. Many among them as they ate must have been thinking of the more famous traveler before them who had as good as dived head first into the forerunners of those very oyster beds, in order to get political instructions from his father down below.

The world was just then being thrilled, something it was still able to be in those days, by the first use of the telegraph, and we like to think that some extravagant amateur with pull at the

Vatican sent Coste or Napoleon III a telegram about his discovery, thereby bringing about the great turning point in the fate of the oyster, not only on the coasts of France but on those of Holland and the British Isles as well. In the nick of time, by that one piece of initiative, for without Coste's study trial and error would have taken too long, Ostrea edulis was saved from extinction, or what would have amounted to that for practical purposes.

It was just in time in another way too; something, some final pout from the old gods, has done away with the business since then. The lake was and is in an old crater and one report has it that this too eventually blew up, but the lake still exists and the guidebook of Naples merely mentions oysters there in the past tense, without explanation. Coste reports that early in his century a burst of sulphuric emanations had killed all the oysters and new ones had had to be brought in, so perhaps the final blow was of that sort. A few more details come from an American visitor later. Anyhow, it can't have compared with the drama at Lake Lucrinus. Oh, Virgil, if only you were here to tell of that! It was preceded by a series of earthquakes over a number of years and the gradual emergence of a beach from there to Pozzuoli. The quakes began to be more frequent around 1536, reaching twenty on the day of September 28, 1538, when many buildings collapsed. It was only the beginning. It is on the following night, our flesh crawling, that we see the Sibyl rising to utter again, after brooding wide-eyed in her cave, in the pose we know from Michelangelo, for fifteen hundred years. Down, Cerberus, good dog, down! Ye gods, what is she going to say?

The sea withdrew two hundred yards from the beach, the earth split open before her cave and all hell at last broke loose. Forty-eight hours later the mountain was there, and only a fraction of the lake, which you can visit now, was left. Ashes and flame and flying rocks are all we are told about, but we know better. The boatman with his boat, awake but not operating, long since struck into a paralysis of torpor by the world's forgetting; the hopeless mobs of shades never released from their attitude of hope; Hunger, War, Disease, mad Discord with snaky bloodstained hair and the other dread apparitions still posturing away to an empty house while the new cast plays down the street; old Hecate herself and all her gang—all were blown sky-high and freed at last, and what the Sibyl said before she too dissolved once and for all nobody ever heard, the racket was so terrible. The rocks and hissing and roaring were only half of it. What was really ear-splitting was the clatter of the oyster shells, century upon century's deposit of them flying through the flaming night, to form what surely must be, if anybody should dig into it, one of the world's most exclusive kitchen middens.

It is not unlikely, since they were members of the French Institute during the same years and were both men of ranging mind and learning, that Coste had at least some slight acquaintance with Hersart de la Villemarqué, who was eight years younger. Though their fields of work had nothing in common, there are some haunting parallels in the lives and mental make-up of the two, making almost a twin image of their large role in this story. To be sure Coste was neither a Breton nor as far as we know a liar; La Villemarqué was born and lived most of his life near Quimperlé. But if poems could be weighed against oysters it might appear that the two were of about equal measure in the total reality of, for one place, Locmariaquer.

What La Villemarqué gave to Brittany was its voice. That he did so by more or less spurious means seems now to be beyond question, but that doesn't close the door on him either as a poet or as a poet of Brittany. Indeed the brickbats fly hardest precisely on the issue of his having been too good a poet. Taken all together as a single epic the ballads of the *Barzaz-Breiz* might almost rank with another obscure case, the work of a bard named Homer, in George Sand's opinion, and she was far from alone in it. The volume, first published in 1839 when La Villemarqué was twenty-four years old, never had any vast circulation but brought its perpetrator, not to call him its author since he claimed otherwise, considerable honor and prestige. He was belaureled by the French Academy, adored and exegesized in Germany, greeted with worshipful acclaim in some quarters in Brittany; there were not many Breton scholars, or Celtic scholars anywhere, qualified to look such a glorious gift horse in the mouth, and some of those were reluctant to do so. However, the storm eventually gathered.

The songs, mostly ballads, some more on the order of hymns or chants, range in attribution from the 6th Century through the 18th, with a preponderance about halfway between. They are divided into three categories: mythological and historical, of fêtes and love, and religious. In wordage the greater part of the volume is made up of an introduction and detailed notes, certainly rather cavalier by the standards of present-day research but fascinating in their range of speculation and the kind of gentleman-scholar erudition now generally lost to the world. The whole notion of "folklore" with its terrible thickets of myth, history and art forms, now taken over by regional propagandists on one hand and the universities on the other, was brand new, and La Villemarqué brought to it, along with vast reading in quite a few languages, the freshness as well as the rashness of a very young man in love—in love with the poetry of a place. The notes also include specific sources for the collection: "I first heard this version from a beggar woman who came to the door . . ."; "My mother heard this sung by a blind man, and later by a young peasant in the parish of . . ." Many of the singers, all of them of the people and illiterate, are named. In a few cases La Villemarqué stated that he had done some editing as between different versions of the same song; he confessed nowhere to having added anything, still less to having made up anything out of whole cloth. The songs, in short, were to be taken as a bona fide and mostly quite precisely dated heritage from the whole past of Brittany since the 6th Century.

This would seem to indicate that his whole ambition, the drive of his self-esteem, was as a literary "archaeologist" as the word was then used, and that for some peculiar reason, possibly touching on social station and love of mother along with whatever else it might be, he really and truly did not want to think of himself as a poet. A passion for Brittany, for instance, sharpened and brought into perhaps shocking focus by some youthful years spent in Paris. Also, vanity takes some queer forms as we all know.

Mazes of scholarship, cauldrons of rage have surrounded the matter ever since. The worst attacks in his lifetime were from literary Bretons; other Bretons, including Ernest Renan, author of the *Life of Jesus* and a native of Tréguier, were less openly belligerent but no less skeptical. At present that side of the case would seem to have had the final treatment, in a lifetime work published in 1959 by another compatriot, Francis Gourvil, a book as indignant as it is thorough and no doubt on most points unanswerable. A grim job, evidently undertaken out of as great a love of Brittany as La Villemarqué's, and for which no stone or village birth registry or scrap of reference of any sort was left unturned to expose the subject as fraud. Some of the people given as sources seem not to have existed; there are glaring discrepancies concerning others; La Villemarqué's command of Breton was shaky and his Welsh worse, his historical associations sometimes wrong when not preposterous, his dating of the songs altogether unreliable; and so on, with some points the learned might debate, as that popular sagas cannot possibly be transmitted orally through a number of centuries without essential changes, and that they never clearly correspond in the manner claimed by La Villemarqué to actual historical events. That is not the view of some eminent savants of Norse and Icelandic. Where the critic is most damning is in the comparison of the *Barzaz-Breiz* with other collections of popular Breton songs also made in the 19th Century. None had the quality, the finish and elegance and completeness of this one. The best the other seekers of popular folk material could come up with was poetic odds and ends, usually fragments rather than complete ballads, for the most part both straightforward in their narrative and rather coarse.

So, he amended, he improved; we won't linger on how much. A curious character in the story is a certain Abbé Henri of Quimperlé, La Villemarqué's mentor, who seems to have straightened out his Breton for him (not enough, it appears) and to have given an occasional *coup de peigne*, stroke of the comb, to the disheveled raw material; and it may or may not be that the two of them engaged for a time in a secretive campaign, or plot, to foist their "popular" songs off on the plebeian singers of the time, especially the blind who were still the main carriers of the bardic tradition; i.e. they were not so much recording a folk art as trying to rekindle one. It could be, along with other part-truths; if so, the effort failed. The songs as we have them are not now being sung at popular gatherings and probably never were. Yet this much is true. For a great many of the ballads, if not quite all, there do or did at the time exist equivalents in the popular repertory, as recorded by people beyond suspicion of being themselves poets, and La Villemarqué's own authorship varied greatly in degree, from very little to very much—or as Gourvil has it for one or two of the earliest songs, total. In that category would be the beautiful "Les Séries," the first in the book, a lesson in the form of a dialogue between a druid and a child, on the mystical significance of numbers from one through twelve. For this the only source is alleged to be an ancient and much-garbled nonsense chant on the numbers, current in various versions, called "The Frogs' Vespers." It seems the language and everything else is wrong with "Les Séries," and the worst is that it is a remarkable poem and greatly admired by "lovers of mystery." In other words if you admire it, (*a*) that's what you are, and (*b*) it's a pretty bad thing to be. You couldn't be just responding to a superior piece of literature, on grounds known to yourself. Still, it is probably safe to assume that it is not literally "authentic"; whether or not it is in spirit one would have to be an expert in druids to tell.

The trouble, one of the two chief troubles, is that the culprit never recanted, and his silence turned youthful indiscretion, if it was only that, into real dishonesty. Apparently he was aware of having given out false assumptions and impressions, but he let them stand in the later editions of the work, and instead of answering his critics, which he was given plenty of occasion to do, took on the injured-dignity role. He was compared of course, to the fabricator Macpherson who had passed off his *Ossian* as an antiquity, and could perhaps have retorted that there are many degrees and several faces of truth in these matters, as witness the more famous fabrication of a source, by Chaucer. Never mind, the truth that matters now is elsewhere; some strands of it might be found, for instance, in the fiction and drama of a contemporary Breton writer, Julien Gracq, who also grew up knowing the Fisher King. La Villemarqué, with whatever degree of original innocence and to put it at its worst, had had his muse or demon set off by a certain Arthurian reality among others in his native landscape, without reckoning with the figure of the District Attorney that would come to loom so large over those beautiful shadows and his play with them.

Worse still, even if his versions never appealed either to the illiterate or to the more careful scholars of Brittany, they did so much to so many others that he became blamed for the whole modern Breton nationalist movement, which had rather faded out for a time before. By this effect of his great epic, for that it truly is, he would undoubtedly have been horrified; his worst enemies have not accused him of having any such aim, but only of having made too much of the anti-French theme, while being himself very much a Frenchman as well as a Breton. A straight line has been drawn nevertheless from the *Barzaz-Breiz*, striking as it does all the grand chords of Breton memory, to the fact of collaboration with the Nazis in some Breton quarters in World War II. The "swan from overseas," on the return of Montfort from England, is taken as a case in point. Another is the ballad on the dominant female figure of the same period, all heroics and ferocity. That was Jeanne la Flamme, wife of the first Jean de Montfort and mother of the victor of Auray, heroine of the siege of Hennebont where she personally set the French camp on fire.

Jeanne la Flamme is the most intrepid on the face of the earth, truly!
Jeanne la Flamme set fire to the four corners of the camp;
And the wind spread the conflagration and lit up the black night;
And the tents were burned and the French roasted,
And three thousand of them turned to cinders, and only a hundred escaped.
Now, Jeanne la Flamme smiled the next day at her window,
Looking over the countryside and seeing the camp destroyed,
And smoke rising from the tents all reduced to little heaps of ashes.
Jeanne la Flamme smiled:
"Oh what a fine weed-burning! my God!
My God! what a fine weed-burning! for every grain we will have ten!
The ancients were right when they said there's nothing like Gauls' bones,
Like Gauls' bones, ground up, to make the harvest grow."

To an outsider, not knowing the Breton language, the same impression of the lady and the incident is conveyed by the guide-books and history books, except that they were not written to be sung, or even recited. However it might be true that La Villemarqué used the terms Gauls, Franks and French a little too interchangeably, and that the drama of the old opposition to Paris occasionally ran away with him. A sense of saga, and of mission as a bard, would almost require that it should, although not that it should pretend at the same time to the trappings of the serious historian.

At his death, at the age of eighty, having carried his peculiar burden of unclarity for fifty-six years, the poet was not without friends or honor, and he died not in a poorhouse but in his ancestral home, which his genius had served to festoon with suggestions of the tarnished and the ambiguous. His son wrote a book later that was supposed to vindicate him and left the confusions intact. In life his powers of charm and intellect, and one gathers above all some massive probity there in the aristocratic shadows, had caused some of his detractors to make friends with him again and renounce the attack, although without going back on anything said earlier, and whether through pride or pathos, such terms of friendship seem to have been acceptable to him. At the same time, to mitigate the picture, he had plenty of admirers, in Brittany and the rest of the world, who without being idiots or fanatics of autonomy could write off the human frailty, the mistakes and the tinkering, the achievement being what it is.

That, since the hatchet work has been done, is surely the side of sanity now. The scholars who because the *Barzaz-Breiz* is "discredited" haven't bothered to read it are missing more than an oddity of literature. It is the saga of Brittany sifted through a romantic genius of that period, as the past of Rome, speaking broadly if there is any need to say so, was through Virgil's, or the Wars of the Roses and Rome again too for that matter were in the English case. And the work is perhaps only at the beginning of its life. At least when the dust has settled, the pathos may be more on the side of two other great romantics who arose in Brittany at around the same time, Renan and Chateaubriand, one of whom by the way was quite a liar too. All you have to remember, if scruple dies hard, is that a song attributed to the 6th or 9th Century is a 19th Century poem built up out of certain traditional scraps, which may be more or less genuinely of the 9th or 14th. Also, a rare verse or two may be what La Villemarqué said they were.

FIVE

THE SECOND EMPIRE—time of Eugénie's hats and the great horse races at Longchamps, in fact the creation of Longchamps as well as the Emperor's pet toy the Bois de Boulogne, and Offenbach's music and the Crimean War and Haussman's remaking of Paris and the black disaster of 1870. A while ago the last century and a quarter was looking very long but that mood is passed; it came from thinking about laboratories. Actually, as the oyster teaches us, everything from Peking Man to now is just our own little house, or *home* to use a real estate term, and we ought to be able to move around in it easily enough. Of course there are funny mirrors everywhere, and a lot of things get mislaid, Coste's initials for instance.

He was one of the eminent men of his time in science, with an extraordinary literary gift into the bargain, author, among other and more strictly scientific works, of a masterpiece of reportage, *Voyage of Exploration to the Shores of France and Italy*, in which the bit on Lake Lucrinus appears. Yet his initial, which seemed to have been V., turns up as P. or J. or almost anything and usually doesn't turn up at all. Nobody but European oystermen or scholars of the oyster, a field that is not exactly mobbed, and perhaps some embryologists, seems ever to have heard of him, and few of those know the rest of his name.

It was Jean-Jacques-Marie-Cyprien-Victor. His dates are identical with those of Louis Agassiz and except for a one-year difference in date of birth, Napoleon III, who is the other shining knight in our story and a more tragic one, being remembered by most of the world only for his failures, only too well. Coste was born in the southern department of Hérault, studied medicine at Montpellier, published a big two-volume opus in embryology and a number of shorter works, as on the ovology of the kangaroo, dedicated the last half of his life to the study of fish and other marine and fluvial life, was President of the Académie des Sciences in 1871 and died two years later. We have not been able to discover if he was tall or short, married, easy or hard to get on with, wore glasses, took snuff, was known to his intimates as Jean or Jacques or Vic, or anything else of a personal nature.

We know everything of that sort about Napoleon III, the libraries groan with it, but he is the one who is opaque, and Coste is clear.

"Napoléon le petit," "the poodle Emperor," the dreamer, the opportunist, "the modern Emperor"—a smear of pitch to some, pathos to others was all that was left of his figure with its noble brow and gentlemanly ways, after the débâcle. The errors leading up to Sedan may have been less damaging to renown than the image of him there in the last few days, wandering sick and humiliated at the front only because he is not wanted in Paris either, not trying to give orders but the contrary, only trying to stay out of people's way and get in the way of a bullet, and denied that luck. It was too pat, after the pomp and glitter and real prosperity of his reign, which to the strains of the *Gaîté Parisienne* had carried through every action the curse of the coup d'état of December 1851. He was not monstrous enough to be forgiven, as his uncle might be; in this case it was all too mixed and nebulous, too human, you couldn't tell where to start forgiving, and praise ran into embarrassment. So it was the killing irony of incompetence that stuck; the human sense of theater had perhaps never been so well provided that way by the downfall of a ruler, even to the lingering on of insane Carlotta, the perfect touch in the gruesome opéra comique, lest Napoleon's criminal absurdity in Mexico be forgotten in the grander finale.

To the intellectual view, the lightweight culture of his period would be sinister enough in itself, police state or not. There had been such splash and fashion about it, all up and down the scale, at the opera and the wonderful Exposition of '67 attended by dozens of reigning monarchs and the somewhat ridiculous tableaux vivants at court, where the reigning literary light, who was also the darling of the public, was a perfectly charming fellow named Octave Feuillet. Baudelaire was in the doghouse. So were the Goncourts for a while, and Proudhon, and Flaubert, although we hear of a reading of *Salammbô* at the Tuileries, after which a butler was sent to a restaurant to order a supper for the company *à la Salammbô*. "À la Salammbô, vous comprenez?" "Certainly," said the restaurant lady, "we have many orders for that now." Only they got bits of fried pork in place of the peacocks' tongues. Victor Hugo's being in exile didn't prevent his *Toilers of the Sea* from setting off a vogue of various dishes of octopus.

A razzamatazz society; vogue was everything. Time of the can-can, the courtisane, and the *blague*, that form of wit so different from the bon mot or the jeu d'esprit. The idea was to laugh at everything. The food at court was merely luxurious, not fine—"a good second-class table d'hôte," the elegant Austrian ambassador Prince Metternich called it.

Academic freedom was sorely pinched, the press more or less muzzled; journalists and professors seethed, to Napoleon's own mystification—he thought they should understand that all he wanted was the good of the people; and his relations with the Academy and Institute, of which Coste and La Villemarqué were members, were uncomfortable. Not really more than that; he had no stomach for thinking of himself as a tyrant and didn't try to stop the election of some of his worst opponents; just uncomfortable. Sighed the monarch, "There is a real conspiracy of literary people against my government."

Poor man, for his reputation there was worse coming. Always

the darling of certain sentimentalists of privilege plus personal charm, he was to have an even unkinder rehabilitation later on Admirers of fascism couldn't do much with the big Napoleon but found a very profitable grab-bag in this one, and fished out an ideal jumble of ambiguous goods and bads—the plebiscite for one thing, which has revived him as a devil all over again more recently among De Gaulle's opponents. Actually, by our standards, as an autocrat he was an old fuddy-duddy, however elusive, and far too nice to learn much from, although he seems to have had some good spies and police people working under him, as Zola ragefully shows. Autocracy is less sensitive and more expensive nowadays. No modern dictator could be such a fool as to neglect the army as he did, and concern himself so much, not just in speeches but really, with the peaceful sciences and the living conditions of the people.

There were more than bad novelists, well-dressed ladies and a confusion of ministers around the Emperor. Pasteur was one of the scientists who were on easy terms at the palace. The sovereign's good heart and keen if not spacious intelligence, scarcely operative in matters of administration, were all there in regard to science, industry, social progress. "Scratch the Emperor," Guizot said, "and you will find the political refugee" —the one whose writings in prison had included an economic study on the burning question of sugar production, source of tragic unemployment through the measures of Louis-Philippe. The two facts that emerge most vividly from Coste's collection of reports, most of which were addressed to the Emperor himself and not to the Minister of Agriculture, are the amount of personal attention and support by Napoleon in Coste's tremendous work in fish and oyster culture; and secondly, the degree of social passion, what used to be called idealism before the word went rotten, that they both brought to it. They were going to create, they thought, by an application of scientific methods backed up by legislation, a nearly unlimited source of good food for the people; trout, salmon, oysters didn't have to be luxuries; they would be cheap and plentiful, for everyone.

Over and over, the theme appears in Coste's reports. A wind blows fresh and strong through all his pages, a directness of heart and humanity, which only a wonderful writer could have managed along with such topics as the breeding habits of the grayling. Whether the same quality cost him something as a scientist we can't tell. He became trapped in the practical side of things, with less and less time for thought and the lab, a fact he himself recorded in one small cry of misery. Still, his regret can't have been unmixed. He was being given a chance probably unique in the annals of zoology to try out his theories on a vast scale, with practically unlimited backing by a government, and besides was undeflatable in his sense of the glory of the work for all humanity. The utopian tinge of the time went deep in him, more likely by osmosis than any affiliation, and there was nobody else in France to do the job; nobody else knew enough. So he did it; *they* did it, he and Napoleon III; either one would have failed without the other, and in a sense they failed together, since the more generous reaches of their goal remained a dream. However, they saved the oyster industry.

Coste was doing a great deal else at the same time. Aside from his laboratory at the Collège de France, where a chair had been created for him, he set up and directed an establishment at Huningue in the Jura, for the artificial propagation of fish—"the grandest enterprise of the century on living nature," made possible by an "unprecedented" allocation by the government. The procedure had been started in Germany in the 18th Century, but Coste was the first to work it out to the point of large-scale efficiency, not to say *grandeur* ("la grandeur de la France") although it is the kind of thing that ought to be meant by that. Scientists came from everywhere to study the methods used, and the fertilized eggs helped to restock streams all over Europe, being sent out packed in wet aquatic plants to any government or private breeder who wanted them, "so that the great experiment involving the nutrition of peoples should have a European character." A far cry from the fascist mentality, and the words could as well have been Napoleon's. The Emperor was in on it all, visited Huningue and the lab in Paris . . . For do-it-yourselfers Coste gives clear directions with diagrams, on how to hold the male fish over the pan and so on.

He concerned himself at length with lobster production; fought for administrative and other changes, from the most sweeping down to details of boat tonnage, salaries and personnel, to stop the ruinous decline of both marine and river fish, and on the whole got what he asked for; forced the adoption of a new kind of sardine bait made of salt capelin to be carried by navy ships returning empty from Newfoundland, to save the sardine fisherman from buying imported caviar, or *rogue*, for bait, a "crushing tribute" that took half their earnings ("C'est la misère!"), with special insistence that the inventor of the new bait, a certain Dr. Balestrier of Concarneau, should be made a member of the Legion of Honor; crusaded for a system of loans and other encouragements to "workers of the sea," similar to those in agriculture—never forgetting the basic point: "I have seen the most fertile productive fields devastated by pillage, when not by ill-timed or anarchic exploitation"; "the government can no longer tolerate an abuse which, if prolonged, would end by drying up the source of all production." Or this, in a report on the trout and salmon fisheries of Scotland and Ireland: "Thus, Sire, on the one hand wealth, solely by virtue of surveillance, cultivation, production; on the other"—France—"ruin, because the rules of a rational exploitation are not observed."

Ah, but all that would be over now—the grand illusion of the century tingles on every page; through the glorious march of science reason *would* prevail, and hunger vanish. "What wealth for France! What a teaching for the peoples of the world!" The "domain of the seas" is public property, the responsibility of the state; the pillagers and anarchic exploiters will be checked, and then the only enemy will be apathy and dislike of innovations. "My perseverance will more than match the obstacles that can be raised, just so Your Majesty continues to grant me the privilege of calling on his high intervention every time resistance threatens to compromise the success of the work." That was written in '61. His Majesty, with his troops in an increasingly awkward position in Rome and the Mexican abomination brewing in his and even more in Eugénie's head, along with the Habsburg problem, did grant Coste that privilege, and was taking equally important and lasting measures at the same time for the improvement of sheep stock, girls' education, wines, agricultural credits, and most crucially for oysters, the railroads.

The oyster question came to take precedence in Coste's work only because the crisis was by then absolute; it was a deathbed call, and fish and lobsters, although suffering from the same general social neglect, could wait. But the two finest sections of the *Voyage*, as literature, (second edition, 1861) are not on oysters but on two very ancient industries, both still continuing now with little change since Coste's time. One is the mussel business in the Baie de l'Aiguillon, near La Rochelle, the other the fantastic system of eel- and other fish-breeding in the lagoons of Comacchio, south of Venice. Like the rest of that memorable book, these two reports were written merely to convey some practical information to the government; it is by a formula of personality and style mostly lost to the world now that their technical nature somehow leaves room on the page for tides, weather, human dignity, and the awesome reverberations for the human mind of the habits of fish. As for the ancientness of the two occupations, although Coste was sensitive to a large gamut of meanings, short of religion, in such continuities, he was not the man to be tickled by ancientness per se. What stirred up his soul in the first place in both industries, and allowed him to be incidentally tickled in so grand and contagious a fashion, was their efficiency.

The mussel story starts with an Irishman named Walton, who one night in the 12th Century was shipwrecked with a boatload of sheep on the French coast, a few miles above La Rochelle. With the ingenuity he soon showed it would seem that he could easily have made himself a new boat to go home in if he had wanted to. Instead he decided to make a living catching sea birds in nets attached to pickets, and to his surprise found a far more profitable crop of mussels on the submerged part of the pickets. From this fact, and Walton's exceptional wits and labors, there soon developed an industry that came to cover some square miles of the bay with its *bouchots*, or picket arrangements—at low tide a design of marvelous beauty, and the sole living, as for Locmariaquer its oyster tiles only in this case for eight hundred years now, of three villages, Esnandes, Charron and Marsilly. The peculiar flat-bottom boat, or little barge with sides, that Walton invented to move across the *vase* is still one of the chief instruments of the trade; there is a knack of propelling it with the right leg overboard. The spirit of the three villages seems to be, or to have been, equally peculiar. Coste quotes an earlier writer on the charity and brotherhood that marked the community, its good humor and lack of quarreling, as in the establishments of the Moravian brothers in Germany and North America.

Another beau métier—from the part of the sea in it, also the enviable mixture of the communal and independent, which was a sphere of economics Coste understood very well. The area remains the greatest producer of mussels in France, and whether or not the old spirit quite prevails, one does have a sense of a certain lively serenity, the kind that grows from a right quotient of adventure in life, along the shore there, even at the height of a spring tide with not a picket still less a mussel to be seen. It is like moving out of a block of fog, but might be partly because the mussels, being a very big investment, have so far beaten off the vacation trade. Also the dreary flatness of the country around doesn't prepare you for the lyrical jolt of the place, more disturbing than any cliff.

The economics of Comacchio were appalling, but as it belonged to the Papal States, having been acquired by Clement VIII from Venice in the late 16th Century, and Napoleon was in enough trouble with Rome already, Coste was free to say so only by indirection. The five hundred workers of the lagoons, under the immediate overlordship of Prince Torlonia at the time, lived not only in a state of serfdom but under military discipline as well, with minimal pay, less freedom, and a system of penalties that would seem more suited to a penitentiary. There had evidently been some repercussions from the uprising of 1848, five years earlier; a fortress on the main island had been demolished that year, and Coste reports that as he was leaving, one of the *vallanti* as the fishermen were called, from the *valli* or sections of the lagoon, addressed him "in these words full of melancholy resignation, at a time when the least complaint touches on revolt: 'May God go with you, and may He have brought you among us in order to better our lot.' " One hundred and twenty of the five hundred were policemen, in boats, and even so the overseers counted on half the huge output of fish being stolen every year.

A social set-up more different from that of the Baie de l'Aiguillon, where every fisherman was his own master, could hardly be imagined. Nevertheless, there too Coste could speak of "the esteem in which they hold the dignity of their art," and their pride in what he calls "this anonymous marvel . . . this hydraulic system unique in the world." Built up over many centuries, the system was based on the instincts of certain fish to go up watercourses a little after hatching and return to the sea when adult. Its setting was, still is, a huge swamp or shallow lake 140 miles around, bounded by two rivers, formerly branches of the Po, and separated from the sea by a narrow strip of earth with an opening and port in the middle. The marvel, as engineering and in taking account of every conceivable whim of the fish in question, was in the arrangement of dykes, canals, sluice-gates, inner labyrinths, etc. whereby water from the Adriatic was kept circulating and the fish were drawn in and trapped or let out to spawn as the moment required. It was not the enormity of the construction that awed Coste, as he reported, so much as the extraordinary exercise of reason and practical sense that had gone into it, not out of one brain but continuously since the Middle Ages, and which showed itself also in the kind of sensible regulation he was struggling to have put through in France, not for the benefit of popes or princes. No offshore fishing with small-mesh nets was allowed at hatching time anywhere near the lagoons, or with large-mesh nets except at a certain distance out to sea; and so on.

Occasionally, from freakish heat or some other cause, there would be a great mortality of fish in the lagoons and danger of plague; or it might happen that the take through the gates was so phenomenal the labyrinths couldn't hold it, and the fish would swarm in to die in great heaps on top of one another, while the *vallanti* worked in frenzy to get them to the cannery. Glimpses of heat-haze, of the Greek profiles of women and the perfect muscularity of men for all their hardship, of a canalful of eels in moonlight or lightning, flash through the statistics, culminating in the great scene in the *manufacture*, not literally a cannery since the fish not sold fresh were packed in barrels. The work there was done mostly by women, in one huge raftered and arcaded room with open-hearth spits along one side. Some of the eels bound for the spit had their heads and tails cut off first, to be given to the poor, but these were only the fatter ones, because of the difficulty of getting them on the spit alive. Those of better quality were "doomed to that torture, after receiving one or two slashes to make it easier to bend them," for they were strung in a series of loops. Coste states that the practice dated from Roman times, and that the dexterity of the spit-threaders astonished everyone who saw it. The grease was caught in troughs and used for lighting lamps and for frying.

An alternative process was a kind of pickling in one of several acid combinations, such as grey salt mixed with a special imported vinegar, the vinegar of nearby Ferrara being too weak for the purpose. Here too the best made out the worst, as a matter of necessity. "Here, as on the spit, this species is doomed to suffer a long and agonizing death. It is pity to see them drinking and spitting out the burning liquor, exhausting themselves in useless efforts to escape by flight from its scorching bitterness, and in their impotence writhing to the surface as though to lift themselves above that pit, where the length of their anguish is the very measure of success in the operation." If they were dead before immersion, their entrails would not absorb enough of the liquid and would putrefy.

The Fusaro report, the one that shook the world so to speak, is rather flat after that, except on Sergius Orata. One wonders if Coste may not have had a sneaking preference for things with fins, a slightly higher form of life, and had to force himself a little in his mighty labors and enthusiasm in behalf of the oyster. What he saw was modest enough. The dark, volcanic, squdgy-bottomed lake, one to two meters deep, was sprinkled all around with two kinds of collectors. One consisted of a mound of stones surrounded by pickets, on which adult oysters had been placed to form a little artificial bank, and where it appears the new oysters were allowed to stay for their full period of growth; in other words, there would be several generations on the collectors at the same time. The other was made up of a line of pickets some distance apart with a rope along the top and bundles of fagots hanging from it in the water.

A modification of this hanging-brush system is used now in the Étang de Thau on the French Mediterranean, back of Sète, and very pretty it is, but it is not for seed collection and the oysters are mostly Portuguese. Aside from that one spot, the techniques Coste studied at Lake Fusaro were quite different from those eventually worked out in France, where the trick was to find some adaptation of methods and materials that would work on the scale required and in less quiet waters. However, they were the point of departure, and the main source of confidence, in that key moment, that reason might after all have a weapon against the forces of pillage and "anarchic exploitation."

The procedure was so novel, there was no established word for it in French. Coste spelled it at first with two *o*'s, *ostréocul-*

ture. The raisers of oysters in parks were still called merely fishermen.

So began a whole new series of trials and errors, by Coste himself and others, based on his Italian trip, while two hundred thousand people died of cholera in the Crimean War, not to mention the insane inadequacy of French supplies in that affair, and while Paris prepared to stun the world at its glorious Universal Exposition of 1855, with its triumphs in electricity, agriculture, industrial machinery of all sorts. We don't hear of a fish or oyster exhibit but perhaps there was one.

Fagots were tried in various patterns, and did attract many oyster babies but also allowed a great many to slip through, and the fagots were apt to be carried away by currents. Coste then designed a huge collector of planks set on beams, like a platform for amateur theatricals, one surface of which was to be roughened by scratching or covered with shell fragments and gravel stuck on with tar, to give maximum footing to the oyster specks; a further idea was to tie branches on top of all that. Something of the sort had been tried in Cancale with success, but the contraption was cumbrous and expensive, and although the little oysters seemed to like it, it was not easy to get them off. Meanwhile different kinds, sizes and arrangements of tiles were being tried, some in slanting rows on wooden frames set on legs in the water, others in a series of pitched rows like a gable, none of which were stable on soft bottoms. At Arcachon a complicated breeding box was developed, perforated and with a set of wire-mesh trays inside, the whole resting on beams. An entirely different approach was the "closed basin" technique, which after long modifications did come to be used in Holland and the Isle of Wight. The idea was to place a lot of "mother" oysters in a basin and have them breed there, but it was found that they soon lost their generative power in these conditions, for lack of a steady flow of water and constant replacement of food particles, if not for more obscure reasons. They were not happy.

A mason and part-time oysterman on the Ile de Ré, after cleaning an oyster basin with care, found seed on the whitewashed side walls and none on the bottom, and so arrived at the first inkling of the value of lime—actually only a matter of clean surface but thought for a long time to be nutritional. Another oysterman at Pénerf, near Auray, carried on an experiment to find out when collectors should be placed. He sampled oysters at every big tide from April on, and discovered that when the "milk" turned bright blue, the seed was beginning to fix on his latest collectors; those put out too soon had become useless. Such were the private efforts percolating around the moribund grounds, all up and down the coast.

Coste's own first labor in the field, a colossal one not only for his own work and daring but in the amount of money and other support given him by the government, was in the long camel's-tongue bay of Saint-Brieuc, site of the first saintly immigrations and one of the strangest indentations in Brittany. On the face of it, it doesn't appear a very likely place for the business—cold, turbulent, sandy; and in fact the effort eventually came to grief and vanished, nobody nowadays seems to know quite why. However, there had been natural beds there, and in the first year Coste's fagot arrangements, under strict supervision by naval personnel according to his minute specifications, brought marvelous results. Euphoria ran high. It seemed there was scarcely a spot of French coast where the oyster population, given the same treatment, could not be multiplied more or less ad infinitum, to be greeted with grateful shouts by all mankind. One spot Coste had in mind where this failed miserably to occur was the Étang de Thau, where seed is all imported; no progeny worth mentioning has ever appeared. Arcachon had been crying for help for several years before he could finally be spared to go there, late in the decade, and there he had his most enduring success, setting up two model "farms" that did more than anything else to establish the industry on its new basis.

One big decisive step in the whole process, which set the pattern of oystering in the Morbihan to this day, was taken by a Monsieur Eugène Leroux at La Trinité, around the point from Locmariaquer. In 1866 he sent for a load of tiles from Nantes, of a kind already experimented with at Arcachon, convex, 33 centimeters long, and set them out in bundles on his oyster grounds. His two young sons, as the story goes, would go out at low tide every day in the *vives eaux* to brush the silt off the tiles. As there were quite a few thousand of them, this chore gives one to think more about the rearing of children than of oysters. It appears that the boys' hearts were in it as much as their father's, and it is not hard to imagine some of their conversations when they had trudged out through the slime, over the gleaming flats, and found half their hard-tended tiles knocked over and ruined. It was discouraging; the arrangement was too rickety. At last E. Leroux, whether *père* or *fils* is not clear so our imagination is left hanging at this point, not only hit on the notion of drilling little holes toward each end of every tile, but what's more had the remarkable patience to do it. Come to think of it, it must have been the boy. Then he piled them in groups of twelve, crisscross in pairs, tied together at the top by a wire strung through the holes, and set them in the mud on stakes. The stakes, about four feet long, were and still are poked through the space in the center and notched to catch the twist of wire at the top.

These stacks, known first as mushrooms or bouquets and now only as bouquets, quickly caught on in the region and have remained standard in the breeding of all Armoricaines. The process of dipping the bouquets in a lime solution soon followed, after a time when the tiles were dipped one by one, which made for some damage to the lime in tying. Monsieur Leroux is reported to have placed his tiles near the natural beds but to have put other parent oysters around them too as a precaution.

Another kind of labor, too vast to think of with comfort, was carried out meanwhile on most of the shoreline in the Auray neighborhood. The mud was too deep everywhere; millions of tons of material had to be brought in to harden the bottom, to make the parks possible.

THERE were plenty of problems left. For one thing, nobody knew exactly what conditions were best for collecting seed. People assumed that the larvae would fix more easily in still waters, and discovered only well up in the 60's that fairly strong currents were necessary. Another question was how long to leave the babies on the tiles. Some thought till the second year, but this involved owning two sets of tiles, a big investment, also the oysters were found to crowd each other as they grew; finally February to April was settled on as the time for *détroquage*. Then there was the cost as well as the composition of both tiles and chalk to consider. The cost was prohibitive for small oystermen in the beginning, and it was a big day in the business when a tile manufacturer announced, at one of the annual gatherings of oystermen in the Hôtel du Pavillon in Auray, that by mixing the famous *vase* of these shores with clay he had succeeded in making a cheap tile. Then natural hazard took a particularly nasty turn, coinciding with the war of 1870; many people died in the terrible cold of that winter, and so did some billions of oysters. Shortly after that Monsieur Leroux alone lost a million or so from crabs.

The ravage from that source can be nearly as much to older oysters as to the young; they can be smothered in the beds by the sand and silt turned up by the crabs, as well as by a kind of marine worm with the pretty name Polydora, frequent destroyer of oyster beds throughout the world. For the seed, after removal from the tiles, a protective box or *caisse* of wire mesh used at Arcachon was introduced into the Morbihan in the 70's. The little oysters at that stage have a very fragile shell and the circumference of a dime more or less, depending on the individual. Their mortality is high at best, and oystermen say that if they were simply scattered on these grounds, as is done in some deep-water cultivation, there would be none left at all in two weeks. For their adolescence and later another kind of enclosure against crabs was developed, the parks, fifty to seventy-five yards square, being surrounded by fences of wire mesh embedded to a depth of a few inches in the bottom and topped by a horizontal molding of planks, protruding outward. Even so, the war with the crabs goes on, like one of those wars of succession the European schoolchild is always reciting. Oystermen make regular rounds of the beds with traps baited with old fish, and in summer some of the women are called back to work for a day or two now and then, just to go around beating crabs with sticks.

But crabs are only one of a list of enemies, appalling in number and more so in their habits. In the original oyster beds the race seems to have prevailed by sheer quantity, but once that state of innocent horror was past, and man had succeeded where all natural enemies had failed, he had to learn a number of unpleasant things. In fact you wonder, faced with that list, why the pioneers in the business didn't call it quits, and which is more confounding in the human side of the picture after all, stupidity or perseverance. But probably it was like the invention of the airplane, or anything else; if you could know all your troubles at once you would never start, and there is always somebody like the boy who made oyster history drilling all those holes, tackling things as they come along.

Mussels, for instance. They can ruin an oyster bed in no time, hence the law against *mytiliculture* in the Morbihan. One offense by the mussel to the oyster is in being even more prolific and a big eater besides, so that it quickly takes over the food supply. Then it smothers it, huge clusters of mussels managing to clamp themselves with all their little hairs to the shell of a single big oyster.

That is no longer much of a problem in the region, and neither is the little tingle-snail sometimes called the drill or screw borer—the *bigorneau perceur*—plague of oystermen everywhere and particularly disastrous in the Morbihan at one time. It ought to be possible to train this little creature for some civilized work, like sabotage or house-wrecking; a great reservoir of power is being wasted there. It is smaller than a half-grown crabapple, but can stick out from where its tongue ought to be, that is its mouth, an instrument capable of drilling through the shell of an infant oyster in two hours and a full-grown one in six. Then it scoops out the oyster meat, unless something happens to make it let go, in which case the victim sets to work making a new layer of shell on the inside of the hole. In early spring the borers congregate in masses on rocks along the shore, where they lay clusters of eggs, shaded yellow to violet. Obviously it would do no good for one oysterman to get rid of them if his neighbor were not doing the same; in this case all the neighbors, having the same interest in the matter, have done the same, for two or three generations. The eggs are spotted and destroyed before Easter, and the loss from this source is no longer serious.

That is not the case in the United States, with regard either to the drill or the other chief menace to oysters in northerly waters, the starfish. "In Delaware Bay alone it is thought that more than one million dollars' worth of oysters are killed every year by the drills. In Long Island Sound and certain parts of the Chesapeake region the destruction of seed often exceeds 70 per cent." "It is difficult to estimate the total losses due to the ravages of the starfish. It is known, however, that in Connecticut waters alone not less than 500,000 bushels are destroyed annually. Great damages are suffered by the oystermen in Narragansett Bay, Rhode Island, and Buzzards Bay, Massachusetts." (Paul Galtsoff, 1949) Dr. Galtsoff goes on to say that "protection of oyster bottoms in these waters is laborious and costly," which is true to some extent of all deep-water cultivation, but also hinges on the factor of irresponsibility. "Surveys conducted by the U. S. Fish and Wildlife Service in Long Island Sound give ample evidence that abandoned private beds neglected by the owners, and unattended natural beds neglected by the state, are the principal breeding places of starfishes." Since such places are not being done away with in most areas and in fact are steadily increasing along our shores, in spite of heroic efforts by certain oyster companies to keep their beds clean, fighting the pest would seem to be a losing game. It is done with a drag device called a star mop consisting of a lot of rope ends in which the starfish get entangled, to be plunged later in scalding water, or by sprinkling the beds with quicklime, found to be toxic to the starfish and not for some reason to the oyster. At present suction pumps of alarming dimensions and various pesticides are also used, with some success; in general what happened to Sisyphus is more like it.

For Ostrea edulis at least in its youth, and in spite of a lot of similar neglect in some quarters, the great protection from this as from other scourges lies in two facts. One is that the beds are exposed at low tide, which automatically keeps off some enemies, as with our natural "raccoon" oysters on the South Carolina shore; it also means that they can be watched. The other is that "tout le monde fait ça ici." Everybody and his grandfather before him has been out there whacking at starfish; everybody's living and professional pride too depends on it. Not that that is an absolute guarantee. Marennes, where the business is just as personal, has suffered dreadfully on occasion.

From the oyster's own point of view, the starfish is one of

the Daliesque nightmares stalking the misty and sunlit flats in the dry hours and carried in the lovely replenishing folds of the returning tide, always there, allowing no troubled midnight or noon's repose but only one placid unending apprehension of ghastliness, compared to which being cut open in a restaurant and ground alive by human teeth must be considered a fairly happy end. "Mme. Denis will eat your oysters tomorrow," Voltaire wrote in a letter; "I might eat some too provided they were roasted; I feel there is something a trifle barbarous, *un je ne sais quoi de barbare*, in eating such a pretty little animal raw." Leaving out the moot question of roasting, he would have felt civilized enough if he had seen the fate of some of the oysters' relatives. The starfish— But how is it that scientists can describe these things with impunity, without anybody calling them decadent, disgusting, obsessed, modern or even gothic because of it? It must be nothing but the tragic sense, impermissible to a biologist, that puts people in such a pet; the thinnest membrane protects us from our dreams; and some would say we have racial memory, and know more about the sea than we can bear to have recalled. But truth is truth, let the membrane break and the sea pour through if it will. Here follows the choreographic synopsis.

The starfish (does anyone among us remember this?) wraps its arms around the oyster, clamping on with its suckerlike feet, and proceeds, not to kill it quickly which it can't do, but to exhaust it, hour after hour. In that horrible embrace, more silent than that of locked couples drowned in the moment of love, a foredoomed struggle between two muscular structures goes on to the limit of strength, without pause and with hardly a motion visible. But then there is no referee, no other creature of the sea is watching, nobody cares but the oysterman and he isn't there; he could know about it only at low tide anyway. The oyster is alone with its agony and the long hopeless fight that it must engage in, for that is its nature, exactly as if it had a chance; in some circles this is called a moral prerogative. It can't even know what the incubus is but only that it doesn't shift, all it does is pull, with a single unrelenting force on the two halves of the shell. For a long time, resisting hunger and dehydration and the lack of oxygen, the captive keeps its valves shut, but at last weakness sets in. Already dying, it still resists, until suddenly the muscles are limp. The shell gapes a little. The starfish has placed itself in the right position for this moment, with its mouth, which is on the inside of its central disk, against one end of the oyster. It sticks its stomach out through its mouth into the opening; suction and digestion do the rest.

The leading roles would be easy to cast; for music and décor, opportunity obvious, especially as in Breton mythology starfish are the wicked fairies. The good ones, "by the grace of God," are stars in the sky. The only trouble with this ballet is that the oyster has no brain and a nervous system scarcely up to physical, still less metaphysical pain.

The squid attacks in much the same fashion, but there are not many around. More prevalent in the Morbihan but somewhat pleasanter is the horrid-looking *gueule pavée*, the eagle ray, which may be up to four or five feet long and can crack oyster shells with its jaws. Then there are forms of vegetation just as deadly, which must be cleared away continually or they would spread to a suffocating mat. The horrid sponge Cliona works havoc on the covered beds as it does elsewhere in the world, but can't stand being dry so is not much of a menace to the parks.

There are also various diseases of oysters, some still undefined and all more or less mysterious when it comes to prevention, although when they extend over large areas this appears to be due mainly to the carelessness of oystermen in sorting their stock. In the southern United States huge quantities of sick seed, from a protozoan parasite called Nematopsis, have been known to be planted in healthy beds. In 1920–21 the whole race of Ostrea edulis was decimated by a slow-working disease which affects the closing muscle of the oyster. It did not spread to the Portuguese but ruined many oystermen wherever the *plate* was cultivated, in Arcachon, Marennes, Great Britain and Holland, as well as all the coast of Brittany.

But the worst affliction of all, even after so much patient effort by so many people, was still people, just as in the U. S. Or no, not quite like that; there are as yet no industries to contaminate the only known waters of the world in which this species of oyster can be bred in great quantity, and no rival claims to speak of on shore rights, although to judge by Arcachon there probably will be soon, and no big increase of population disposing of more and more garbage etc. except in July and August, and no drastic change of human personality in recent times, imminent though that may be too. The curse of the *ostréiculteurs*, as elsewhere, has been the draggers of public beds, working on a highly competitive grab-as-grab-can basis, with too partial and shadowy an interest in the future to act on it even if they cared to; in other words the same old irresponsibles, as often by circumstance as by character, that we were hearing about in the 18th Century.

The oysters in the private parks are not enough for reproduction. There have to be natural banks in the vicinity, supervised by the state, to provide spawn. In this case the state is represented by *La Marine*, which like political entities the world over is open to various kinds of persuasion, and which besides has a legal commitment to fishermen including the *pêcheurs d'huîtres*—but not the true oystermen, the breeders. This comes under the system of naval recruitment of all fishermen, set up by Colbert under Louis XIV and still in force, involving certain shore rights and privileges in return for compulsory naval service. So the beds went on disappearing all over the area, from Auray to Quiberon, requiring a constant search for new places to put the tiles. Only in the last few years, through the efforts of the little Auray office, a few have begun to be reconstituted, which through an ounce or two of political pressure could be wiped out again.

One wonders if Coste in the last three years of his life, after the débâcle, was seeing ahead to so long a struggle, so small a triumph compared to what he had dreamed; or for that matter if he lived to see all his collectors at Saint-Brieuc washed out, by currents or human indifference, whatever it was. Huningue was to leave no trace either, and his own name very little. As for the state that made it all possible, perhaps he was glad to see it go; perhaps it had become wearing to owe so much to a monarch from whom so much sorrow flowed, and more wearing to know the good side of him. There is no way of telling. We don't know where he was that night after Sedan when Eugénie, after a round alone in a horse-cab to friends' houses where nobody was home, was saved by her American dentist, Dr. Evans, or what he was thinking while across the Channel the ex-Emperor worked to invent a cheap stove to make life easier for the poor in winter. He may have died thinking the Republic would do as much and more for oysters and the rest of it once it had settled in, and without the price tag; or in despair, seeing the regulations coming apart at the seams and the money withheld. He was President of the Académie des Sciences in '71—from eminence? as a stopgap?—but the last actual word we have is ten years earlier, when the shores were going to be teeming with oysters everywhere and nobody would have to be hungry any more. Then there is a postscript, from an American in 1890.

IN 1890 a marine biologist named Bashford Dean, working for the U. S. Fish and Wildlife Service, was sent on a journey of investigation to all the main oyster centers of Europe and wrote two extraordinary reports, one on ostreiculture in France, the other on the rest of Europe. Extraordinary in that they were as good as they were useless; everything has gone on as if they had never been written. They are in the archives, among the publications of the Department of the Interior. Not that our scientists, to be repetitious, aren't as informed and dedicated as any. The question is only whether their hard work and knowledge has any serious relation to the survival of molluscs and fish, and the answer to date is no; it has not. Every year, speaking collectively, we know more and have fewer oysters, and most that we do have have less taste. Conservation has become a battle-cry for everything except, to use another nice word that we ought to be hearing a lot more, aquiculture.

In the post-Coste period when Dean made his trip, experimentation and excitement in the oyster field were at their height in many places in Europe. A dewy rage and radiance still hung over the business; questions were being asked, governments castigated; nothing was too difficult to try. It was the year of publication of John R. Philpots' sprawling opus in England, *Oysters and All About Them*, an all too accurate title; the English government was so remiss, the country's scientists were trying to raise money out of their own pockets for the construction of a marine biological laboratory. The German government was more farsighted and willing to pay but was up against a theoretical obstacle; their distinguished Professor

Möbius couldn't see the oysters for the theory, an invention called biocoenosis or life-balance, which required that the oyster population should not be able to increase in any given spot, because if it did its enemies would also increase, the food supply would give out and all would be back to normal by the next year. He therefore declared impossible what was being done with great success under his nose, in Holland and elsewhere. As Dean observed, playing the classic innocent-American role of the time as the knight-errant of common sense, in a place good for oysters to begin with, the food supply in the sea was more or less infinite; the question was the protection of seed-beds. The Italians had a good marine laboratory in Naples, quite unconnected with a splendid system of oyster cultivation at Taranto, which apparently had gone on undisturbed by science or anything else since before the Empire. Like the Fusaro system and unlike the French, called horizontal because the oysters are spread out over the whole intertidal zone, this was vertical cultivation, in a large sea-fed lake called the Mare Piccolo, with the unique feature of ropes as the supporting material; a single rope fourteen feet long could bear as many as two thousand marketable oysters. Hanging baskets were also used for fast élevage.

Dean visited the Avernus area too and rounded up, there and in general, a great wealth of information on the conditions of pond as against shore cultivation—enough to have reversed the whole sorry American picture if we had let our knight-errant be anything more than an expense to the taxpayer. In some ponds there was a danger of too much fresh water, in others of too much salt; many were slowly ruined by accumulation of sediment. Of what was left of Lake Lucrinus after the great upheaval of the Renaissance, he wrote that the salinity was more easily regulated than in Fusaro and the bottom was better, tufa and sand with less mud, and the hot springs were good for the oysters in winter. But the spatfall was irregular, since the incoming fresh water made it dangerous to open the sluice-gates too long; especially in a warm season the larvae had to fix fast or they would escape to the sea. The Fusaro story was still not quite finished. The industry there had been ruined, according to Dean, by (1) the decomposition of organic matter which had poisoned the water, (2) overcultivation of mussels, and (3) excessive salinity caused by the opening of a second outlet from the lake into the sea. There had been volcanic troubles too, and the water had become too shallow and therefore too hot. At the time, one man had managed to revive the business there, by dredging and opening another fresh-water inlet, so it was some time later that the oysters gave up for good.

The most common experience in pond culture was great success the first year and failure in the second or third. That had been the case without exception in England, owing to excessive salt, sediment, malaeration, the fouling of the water by the decomposition of its inmates, and "general mismanagement financially." In America all efforts had failed in the same way. Dean blamed bad aeration in most cases, and the fact that efforts to renew the water by supply and drainage currents had led to escape of the embryos, as at Lucrinus. There was only one pond, of all the new ones being tried, that seemed to him to promise continued success, and that was at Locmariaquer. He was wrong about it; it came to grief or at any rate was given up like the others, but at the time the tiles in it were averaging 400 spat and sometimes there were the phenomenal number of 3,000 to a tile.

Of that pond, or really lake for it was quite large, what is left is the little body of water below the dyke and the pink mansion, where a few unpretentious fish now whisk through the reflections of the horses. The lake was their field and the other fields around, originally all a salt marsh, converted for oystering as many other places were then by the construction of the dyke and the grassy sea-wall, hardly noticeable now, running along from Françoise's hut. The dyke looks older than that, and so massive, it seems out of scale with any possible profits from oysters, but that only indicates the vast confidence and energy that had swept through the métier. Many similar engineering works in the Auray region date from the same time, all later abandoned as the almost hopelessly muddy shores were gradually hardened, or "macadamized," by even more incredible work. As Dean wrote, there was not a spot along the estuaries there that would not have been written off in America as totally out of the question.

The presiding ghost over it all was Coste, whose name was just beginning to be rehabilitated after twenty years of tragic misappraisal, when nothing was seen but his failures—at Thau, at Saint-Brieuc, nearly everywhere. It seemed he had miscalculated in everything. "His failure," Dean wrote, "he recognized more keenly than did his enemies," and quoting a French acquaintance: "He died, blind, in disgrace, looked upon as almost a charlatan." Then the irony loved by the gods of science and art, of the happy ending. By the time Dean came along Coste's achievement was well recognized. "It was he who furnished the ideas for others to profit by. . . . The industry is a profitable one to the culturist. To the state it returns in rentals a greater revenue yearly than the entire sum expended in the failures of Coste."

For us there is a worse irony in the thought of the man who wrote that, the solid practical American, Dean himself, who was certainly no socialist any more than Coste had been. They had one point of conviction in common, born out of no theory but merely out of the dismal facts and plain necessities of the situation; only there was no government to give Dean any allocation, "unprecedented" or otherwise, with which to try to save the American oyster, nor to provide any of the other necessary conditions, so presumably he was spared a death in failure and disgrace and could have one of petty oblivion, like any other government employee whose reports are not read.

"It will be seen that in those countries alone where government has absolutely preserved supplies of spawning oysters does seed culture flourish." Over and over he made the point, not quite with Coste's mastery of prose but with no less feeling and for the same grimly obvious reasons, that without decisive and large-scale federal action the jig was up. You get to like this man Dean; you could almost cry for him, thinking of it now. He was no wit, just extremely intelligent, and one deadpan sentence he got off has real grandeur: "*We should not expect seed to be abundant where oysters are lacking*." This is truly worthy to be engraved over the portals of the Department of the Interior, along with the quote from Benjamin Franklin or John Stuart Mill or whatever we have up there now. He made just one terrible miscalculation, of a Costian sort. Such action by the federal government, he wrote, "would be heartily seconded by the oystermen themselves."

If he could see Long Island Sound now!—the sadness around the offices of the few remaining oyster companies; the starfish coming up by the million in the suction pipes; the seed all gone and the last erratic harvest taken from the pollution of New Haven harbor; the fabulous amounts of money spent on the newest chemical which it is hoped will not be poisonous to the consumers of a product that in any case continues to dwindle; the wonderful jars in the Milford lab each containing a million healthy larvae, both O. edulis and C. virginica, for which the only foreseeable future is to be thrown out to the starfish and the drills; the earnest talk of needing "more control of the environment than we have had" and the rock-headed determination, at least among most of the older generation in the business, that such control should not come from the only possible source, the federal government. No, for Mr. Bashford Dean there has been no happy ending, not yet. Coste was luckier. There is just one place that it might please him to see. In a pond on Fishers Island, of the kind he studied so thoroughly, a private venture in oyster cultivation has just now been started, with great initial success, using Japanese-style rafts from which collectors of styrofoam, asbestos shingle and meshed bags of cultch are vertically suspended. Whether this brave experiment, so new and daring here now, so familiar in many countries seventy-five years ago, will fail as most of those did or survive all the complicated hazards remains to be seen. At the moment it is a small shining and single hope in the area. Labor costs are not the key; if collectors are properly coated, with lime or an equivalent, machines are quite feasible for removal of seed, far more so for the virginica than for the oysters of Brittany; what has become exorbitantly expensive is the lack of such cultivation. But it will not be simple, we have started very late, and some might say that nature is angry with us.

As for introducing the *plate* into this country, that would be even less simple, though perhaps not impossible. It was tried in Maine in the 1950's, on the instigation of our leading scientist in this domain, Dr. Victor Loosanoff, and failed partly for a reason of great interest, if his conjecture holds. The shipment of oysters in the case had come as adults from Holland, but some had presumably gone there as French imports; it seemed to him that the samples were of different races within the species, with their chances of reproduction cut down accordingly.

At least an image arises for our mild scrutiny out of all this, if nothing else, which seems to stand in some truthful conjunction with the pond on Fishers Island. It is of the American named Dean, paid by the American government, thinking as he stands by the pink mansion of the pathos and triumphs of a man named Coste. Before him spreads the lake covering twenty acres, teeming with oysters just as in Coste's wildest dream, in what must have seemed the perfect vindication of that dream and promise for the future in his own country too.

SIX

SHE KNOWS his hours, his comings and goings in and out of the grey quarters he shares with the old curé, behind the church across the street, knows them as though they were a traffic within her own bloodstream, and must feel sometimes that she knows best of all the hour of his death. Far off, for the Abbé is only thirty-five and sturdy enough to throw two drunks at once out of his folklore festival or the weekly boule games he presides over. She worries incessantly over his health, the signs of exhaustion at the top of his cheeks that nobody else seems to notice, the wretched meals he has to cook and clean up himself for the two of them, the curé typically finding excuses every day to let the younger man do all of that, since the death long ago of the housekeeper. They know in Vannes that he is killing himself, having to do almost all of the aged priest's parish work together with his own at the school, but the new housekeeper so often promised remains on paper. Still he will not die for a good long time, yet carries that hour visibly to her like a ring around the moon, wherein her own pale little light dissolves in perfect happiness.

Upstairs over their tiny corner store with the one sign, ALIMENTATION, her father who used to sail all the seas sits crippled with arthritis, no longer able to come downstairs or attend to *le commerce* at all. A year ago she was obliged to give up her just blossoming life as a schoolteacher with the nuns in Quiberon. Her mother needed her, couldn't run the commerce alone and of course wouldn't hear of giving it up although Marie-Jeanne's salary would soon have done as well for them, put together with the pension. And she is an only child. She herself died then, at least to the world, and knew it; and knew it had been only from pride and not love of her, because they owned a store and were not in a class with the oyster-workers, that her parents squeezed and strained to go on sending her to school after the age of fourteen when the other girls went on to the chantiers or at best to the nuns' housekeeping school in Carnac, to learn to be servants in summer. For a Hémon, unthinkable. She is twenty-two now and old, no longer shy little Marie-Jeanne or the beautiful scared novice teacher, for a while almost engaged to . . . All of a sudden she became Mlle. Hémon the pious spinster, and some might say the daughter of a witch. The mother has in fact developed some increasingly fine instruments of psychological torture, whether in the presence of customers or in the grim little family circle upstairs, and what was said to have been a strong resemblance to her cousin the Postmistress has turned to a blotched and beady caricature. Something other than flesh might be suspected to be itching under her dress in certain atavistic spots, and she drinks; not all the time, just sometimes.

But Marie-Jeanne, who takes pains with her short chestnut hair only, almost only out of respectability now, and has no idea that her neat well-spaced features radiate both distinction and a most fetching plea for hugs and rescue addressed to nobody in particular, in the deep reverence of her spirit knows that merely to honor one's parents is not good enough. Her duty is to love them, and with God's help and the Abbé's example before her she most often does. A year ago when regrets, longings, interpretations were running wild in her, she poured them out to him, and of course he counseled her in the only sense possible, but simply and sadly and without any big words needed, especially not "sacrifice," which wouldn't have occurred to him in regard to himself either. Both knew the deeper laws too well for that, so it was only sympathy that mattered, and helped make the delicate bond between them, for only a few wicked tongues to waggle at. Their private conversations, since that one, are mainly about certain school and parish chores she undertakes, and are held in the public gaze, outside the store or in chance meetings around the bourg. She has, of course, thought of becoming a nun, but as the Abbé rather whimsically agreed she somehow lacked the vocation, which a lot of women might have along with a love of God less true than hers.

Thursday in the dozey hour after lunch, when he emerges with his bike, she is dusting chocolate, petit-beurres, postcards and caramel-candy jars at the shelves crossing the window. They exchange a usual, friendly smile past those commodities and across the cobbles, hers a slight sea-ripple, his the genial flash which no amount of village pettiness or other difficulties has ever made perfunctory. The readiness to laugh and enjoy is not the only depth to the man, but is so basic, a Paris Jesuit might mistake it for silliness. He has read Bernanos' *Diary of a Country Priest*, found it shrewd in parts but on the whole not true, didn't worry much over it. If his own thirteen-year tenure in Locmariaquer were to be written it would have to be in the style of historians on royal reigns, as the time of good cheer, by a subtle not to say insidious triumph over the forces of gloom and indignation officially ruling in the person of the old curé.

He is good company, with a fine baggage of jokes, stories, songs although he may be half asleep on his feet sometimes, having slept five hours a night at most for years. His indignation, very unlike that of the curé who only rails at sin, is sorrowful and more secret. Probably only Mlle. Hémon and Sister Sophie, the nurse among the four nuns, who shares the feeling, have seen it bare. The hopelessness of some of the oyster-workers' lives gets him. Himself from an inland town, where his father was a baker and his brother is a cobbler, he has come to know very well the mysterious something, call it bounty, in this métier. He doesn't minimize that, he sees it every day in the children, even the worst of them; the spirit is better than in schools ten miles away. There are no really vicious boys in his classes, none that he completely fails to touch, in one way or another. He knows very well too that a good living can be as bad for people as a bad one; he has seen something of that creeping cheapness in the modern world, over and above the truth of the Gospel in that respect, and is glad to be where he is. Just the same, there are moments when all the distractions he works so hard to give the people, the games, fêtes, picnics, even—God help him—the trips to Lourdes, seem only a dyke thrown up against desperation and he wonders how long it will hold, or at the blackest even whether it should. It is all very well for the owners, big or little; for the others *c'est la misère*, with no future at all, and he is supposed to keep them off the bottle . . .

But such thoughts could lead to the end of everything, and luckily he hasn't got time to dwell on them.

If he lives with a more personal question mark, or sorrow, or possibly anguish, it would have to be infinitely more secret. He is physically strong after all, like all his people, and attractive, of nervous and passionate nature, not the kind to drift through any human experience lukewarmly. But certainly anything that might have to be wrestled with at dawn or midnight would have been answered long ago, before his knowing her, even though the forms and degrees of suffering might be beyond any earlier imagining.

No, there is no question of the outcome, or rather of anything leading to one at all, although in the blotched and canny features of Mme. Hémon at the upstairs window, also to be greeted, you could read all sorts of suppositions. She gets off a correct nod and smile, rather ghoulish in effect, as though she had just looked up from pinching a baby. What she has actually been doing up there, for years, is to weave a mesh of guilt and ridicule around her husband because of his disease, but he is not a very good-hearted character either and may have asked for it. On the whole the Abbé is sorry for her and doesn't think ill of her, in spite of her evil thoughts and her spying, her almost supernatural trait of being there watching and listening when one could have sworn she was somewhere else. She is, of course, watching over her daughter's virtue, ready to pounce on any sign of strain in the girl and laying little traps to bring on what she is watching for, but the main result has been only to increase her own unpopularity. She does her best to be pleasant to customers, but with three other such food stores in town would soon be out of business if it weren't for Marie-Jeanne.

Not that the old woman's primary assumption is wrong, in one sense. Here, but for the will of God and whatever else put him into the soutane, are a man and woman born for one another body and soul; anybody could see that. The tide between there and Port-Navalo would not be stronger than their single current, if it were allowed. What public opinion calmly knows in this case, and Mme. Hémon with all her snooping misses, is the strength and rightness of spirit in them, rarer than all the rest. It baffles her nearly beyond endurance that after that brief three-way encounter, she upstairs, the two of them below, Marie-Jeanne decides to go fishing for palourdes because it is such a nice day and Papa has been wanting some, and sets off for the beach on her motorbike, with a girl friend, looking quite serene; and that the Abbé can exchange a word with the butcher and his wife up the street, prime representatives of the enemy to the old woman so she feels the word is about her, and go normally on about his errand—a sad one today, as everyone knows.

At the sound of the clacker rigged over the store door she comes down for the customer, only a little boy after a fruit drink, and with quick breath searches the dim-lit clutter for some tell-tale sign. The piece of paper she snatches from the floor is a candy-wrapper; the tears she thought her prize are from a leaking vinegar cork. The little boy puts down his coin and gets out fast.

All, all eludes her. She feels herself the victim of vast injustice, is suddenly licked by fatigue and ready to cry herself. How unfair everything is! But she has no idea how much is really eluding her, so is able to pull herself together.

The Abbé stops briefly at the school, the yard a bare gravelly enclosure with a shed for shelter across one end and the two outside urinals smelling in one corner, the two barnlike schoolrooms with their tough rows of ink-stained desks, French flag and crucifix, strictly swept and put in order by the boys Wednesday evening. The state exams are coming up, there has been a lot of Thursday tutoring but sinewy old mustached Mademoiselle, his teaching colleague, is seeing to it this afternoon. He pedals on out past the gypsies and the Table des Marchands, in gleam of sea and lilt of clouds, thinking of the funeral tomorrow and what he will say to the young widow. The man, twenty-six and an oyster-worker, found in a ditch last night with his motorbike, had no doubt been drunk as usual, but the Abbé will not be the one to say it and doesn't even need to think it; there was some evidence of a blowout and he was not a bad young man. There are only two sets of people the Abbé will ever call *méchants*—the Germans, that is the ones who were operating around there when he was at the seminary in Auray, and the town's rival football team, who have been trying to steal his best players shamefully. If the poor fellow had been killed at the chantier there would be remuneration; now his wife will have to go back into the oysters, and the children are too young.

He is passing the after-lunch stream of women on bikes going the other way, back to the yards, some taking their children to work too since it is Thursday, and yesterday's death lies on their greetings to him. Not many of them from those outlying parts go to church regularly any more although they send the children; hence some of the curé's vituperation and alienation from them, but that is not in the Abbé's role, or his nature. He finds the fact unfortunate but doesn't dislike the people because of it; Yvette, for instance, with her grand smile edged today with the common trouble, for she has just been to see the widow and stops to speak to him about it, or lusty weather-beaten Mme. Aurogné, nearing fifty and full of jokes, taking her two youngest to work with her. And Jean-Pierre? Ah, that's no joke; she is worried, he disappeared a little while ago and now she has to go and doesn't like to leave him wandering in the fields . . .

Later, on leaving the stricken house, the Abbé finds him masturbating by a hedge nearby. His features are blobbering and out of kilter, with one eye vacant and much bigger than the other, but he is not quite an idiot, and his mother's earlier fears of sex violence haven't been realized. Something is preying on him, he manages to say it finally; it is that he is twenty-six too, and now that other one he has known all his life is dead and how can that be, when they were both twenty-six. His head sways like a bear's under such a weight of thought and effort of speech, and then he solves it all in a gesture the Abbé is not quick enough to stop. His huge hands are into the thorn hedge over his head and with one grab, spurting blood from all his fingers, he wrests out a bouquet, and whether from the fragrant little blossoms in it or from his own wounds, is suddenly at peace. And so, trailing vines and specks of blood from his hands, in a shower of larksong, lumbers off to join his mother at the yard, on the Abbé's suggestion; there are a few simple jobs he can do there, though not for long at a time.

Watching him go the Abbé has an impulse, peculiar for that time of day, to kneel in prayer, but merely crosses himself instead, which is not like him either, at such a moment. Then off to the afternoon's football, the schoolboys today, not the older working-men's team he has Sundays, in the field back of the big dolmen. Probably it is time he stopped playing on the team himself, but he still enjoys it and the men seem to want him.

Tomorrow the funeral, Saturday the boys' exams all day in Auray, Sunday the fête of Saint-Pierre with the bonfire on the beach. Then he must start lining up performers for the folklore festival, and planning the spring excursions, perhaps one to the island of Bréhat this year, and for the children a visit to the shipyards at Saint-Nazaire. He must be sure there is no slip-up about the busses, as there was last time . . . He is only of medium height, thick-haired, not noble-browed or with any other one feature of marked distinction; the hazel eyes are only markedly honest, not beautiful nor with any of the depths known as soulful. A face speaking of no repose, none of the chance idle reflection that tells such truths of character, this one seeming molded only by the two poles of total participation and deep sleep. It is a wholesome, strong and sociable face, quick and genuine in laughter; take away the soutane and he could be mistaken easily for his brother the cobbler or his father the baker in his younger years. If the mysterious property that emanates from him is some version of holiness, as it must be, it comes with no more of an air about it than the candy he always has in his pockets for children or a home remedy for rheumatism. Yet there is a trick to the image, rather like that played sometimes by the field of three thousand menhirs in Carnac.

In certain changes of light, or even without, this modest priest will seem to appear striding over the tallest hedges across the fields, with giant steps and the mien of a slayer of dragons.

THERE are plenty of Frenchmen, and some educated ones, who have never heard of the Morbihan by its department name. You wouldn't think so in July and August when all forty million of them seem to be packed into a little fringe a hundred yards wide along its coast, but it turns out the same fringe or worse has sprouted everywhere on the French coast unless there is something wrong, such as certain death from riptides or rocks.

They should have heard of it. It is the only one of the five departments of Brittany with a Breton name, and everybody knows about Vannes and Caesar's defeat of the Veneti, in 56 B.C. That is still rather a sore point in the region, because the Veneti gave him no end of trouble before they were licked, obliging him to build a whole new set of ships, and Caesar was quite nasty about it afterwards. It was only the wind dying down that day that did them in; they couldn't get out of the way of the Roman galleys. It was also only a sea or air mile from Locmariaquer, across the strait at Saint-Gildas-de-Rhuys, next door to Port-Navalo, that Abelard lived in a monastery after his trouble over Heloise. He had more trouble there. The monks had become corrupt and wanted to kill him, so he finally escaped out a back window. The story doesn't say how he proceeded from there, probably on foot. Saint Gildas had it out with the devil in the same place by jumping several miles to an island, in the tradition of Breton saints, and leaving his footprint on the rock forever after. Of course Abelard wasn't a saint, but then neither was Tristram and he made one of those jumps, after escaping out a back window, and his footprint is still to be seen. But that is a little up the coast, in Finistère.

Finistère is the eye-catching department of the five, that and the Côtes-du-Nord, ending at the Normandy line, one side or the other of Mont-Saint-Michel depending on the course of the river there. There is a poem about that: "Le Couesnon dans sa folie,/ A mis le Mont en Normandie." However, since the river might come to its senses and change again, Mont-Saint-Michel is still in the guidebook of Brittany.

The Morbihan can't compete with that, nor with the other spectacular rocks of those coasts. Its natural beauties are mild, and with one exception more on the lyrical side; relatively speaking, it is poor in sculpture and poor in saints. It couldn't be Brittany and not have a great many granite crosses by the roadsides with or without an adumbration of Christ on them, and a fair number of saints, who when they weren't leaping to a rock were usually sailing somewhere in a granite tub. There is only one Breton saint, a pregnant lady at that, who is said to have traveled in a sealed barrel instead, in the manner of Danaë with the infant Perseus—one of many such classical infusions. There is also only one Breton saint, out of the enormous list, who was actually canonized, but that is not at all the reason for his being the most popular; it is not in the nature of Breton worship to make that distinction. This is Saint Yves, a lawyer of the 13th Century about whom the poem goes:

Sanctus Yvos erat Brito
Advocatus sed non latro
Res miranda populo.

("Saint Yves was a Breton, a lawyer but not a thief, a fact of wonder to the people.") His home town was Tréguier on the north coast and his cult is most fervent there, but he appears in churches all over Brittany, always between figures of the poor man and the rich man and often as a member in full standing of the cardinal virtues. Only Christ Himself and the cult of the dead, ornately but somewhat erratically Christianized, are as prevalent.

The saints of these parts, like the general run of their many colleagues all over the peninsula, are mostly older, dating from the time of big missionary traffic from England and Ireland, and of dragons. That was 5th to 7th Centuries, making them more or less contemporary with King Arthur and Merlin and the enchanted wood. Some have been miraculous a while and been forgotten, and many are remembered now only because of some lonely chapel dedicated to them, like the one to little Saint Avoye near Le Bono, decorated by the fairly common Breton motif of crocodile heads carved at the ends of the oak ceiling beams—perhaps a sailors' version of the old dragon, with something more to it, possibly the monstrous crocodile that in one Celtic version caused the Flood, although the beams may end with boars' heads instead. One thinks of Avoye as a girl, and little, because there is that kind of pathos to the place, and her stone boat which is there wouldn't hold a peck of peas. But a lot of work and money went into her handsome chapel a few hundred years ago, and perhaps she was really strong and excessive. There was one like that not far away, who cared so much about her chastity, after being forced into marriage she prayed that the child that was in her would never see the light of day; so her son, Hervé, was born blind but made up for it by becoming a much more important saint than his mother, with a special gift for bemusing wolves. There is Saint Gildas, the jumper, and Saint Cordély who still has a day for blessing farm animals once a year in Carnac, and Saint Cado whose tiny chapel, on the tiniest of islands, is one of the prettiest pieces of small-scale architecture anywhere—a good deal too pretty in its restoration as compared to such a piece of genuine fantasy as at Saint-Gonéry, for instance, on the north coast. The 19th Century granite calvary outside it is pleasant enough, nothing more. The Morbihan has only one of the great 16th Century Breton calvaries, with their heartbreaking, lichen-crusted scenes of the Passion, and that is far away inland, at a speck of a village called Guéhenno.

Saint Cado started as a prince of Clairmorgan in Wales and ended as Bishop of Benevento near Naples, but had a typical tussle with the devil in Brittany in between. His island is only a few yards off the mainland near the swirling mouth of the Étel, one of the places oysters are raised in their second and third year, after they have left the Auray region. The currents are too strong for seed oysters and were too strong for the saint when he was trying to build a bridge over to his church, only in that case the devil was responsible. Every night he undid what Cado had built during the day, so the saint came to parleying with him and the devil agreed to leave the bridge alone in exchange for the first living creature that would cross it after it was done. No Protestant church and no American mother would approve of what Cado then did. He accepted the bargain but hid a black cat under his cloak, which he sent across the bridge as soon as the last stone was in, and the devil could only acknowledge that "once more" he had been outwitted by a good man. It is a peculiar fact of Brittany that the devil is always being outwitted, by practically anybody; or in the broad-jump he is always the one who falls short. The dragons have a strange trait too; they hardly ever have to be slain; the saint merely asks them to leave and they go shambling obediently off into the sea. But evil wasn't always such a booby. Pôtr Penn er Lo, who wanders around Quiberon and the Côte Sauvage on dark nights, has wrecked many ships and will lead lost travelers out into the wicked *vase*, although if he has a mind to he may steer a ship safely home out of danger too. Capricious Celt, he has it all over the righteous and the damned alike, but then he was here before all that came in, and stayed. The climate suits him.

The regular saints of the church get rather short shrift with all this competition, but what is most astonishingly dim in Brittany, as compared to France proper or any Latin country, and a lot of talk to the contrary notwithstanding, is the Madonna, and the female principle in general. The Jesus of Italy is a baby; in Brittany He is Christ Crucified; the mother figures very little, either by the cradle or the tomb, the Pietà too being a theme of no great depth or frequency here. Of course it appears, most wonderfully in the sculptured calvary at Guimiliau, and there are quite a lot of Notre-Dame churches which come in for due devotion. It would be easy to be a little too pat about the church's undying struggle to impose the proper forms in this land strewn with symbols of sun worship. And since it is the cardinal cliché of the place that Bretons are unusually devoted to Mary, we have to assume there is evidence for it. Nevertheless, if you have moved from Latin Catholicism into this, you become aware after a while of some huge, obscure change of religious atmosphere; things are not on the same axis; a different set of emotions are called up by the objects you see. It is partly that the all-embracing Mother, the Mamma (in Breton, Mamm), you were used to is not there. The roadside shrine to Mary is rare, and always cheap modern work. But the granite cross at the crossroads is one of the famous facts of Brittany, nobody knows how many there are, and the native genius for sculpture had its universal expression in the figure of Christ. The wonderful painted wooden figures are of almost anybody but Mary, and many of the sacred springs that the region is also famous for, which would have stirred up virginal connotations in any Latin breast, were taken over after Christianity came in by local male saints. As a matter of fact, where female nomenclature does appear in these matters you can't always trust it. Among other instances, little Saint Avoye about whom we were just musing appears actually to have been a man, whose attributes at some tardy point got amalgamated with the name and story of a daughter of a king of Sicily.

As for the holy grandmother, that is more complicated. She was really a Breton, so the story goes, born somewhere near Brest, and it was only through an accident of travel that she ended up in Palestine—a rather consequential trip, but the Breton mind deals easily with the far and wide. Anyway, she was for a long time the great patroness of Brittany, most notably associated with the 9th Century hero Morvan, the victor in several battles against Charlemagne's successor, and subject of rather a muddle from La Villemarqué, who calls him Morvan Lez-Breiz and makes him the prototype of the Welsh Peredur and the world's Parzival-Perceval. Anne is his "dear Mother," invoked before battle and beside him in the fray very much in the fashion of Homer's Athena, and not to blame for his death in 818, which caused a temporary submission of Brittany to the Franks; on the contrary, she saw him through a miraculous resuscitation and after seven years of penance had him back in arms. (The historical version is that another national hero, Guiomarch, took over that time.) Morvan promised Saint Anne a chapel, along with a "belt of wax" long enough to go three times around her church with its entire grounds and cemetery, a banner of velvet and white satin with a support of polished ivory, and seven silver bells to sing gaily over her head day and night. A certain 13th Century chapel, whether by direct lineage or not, took on that association, and would seem to belie the assertion of one Breton scholar, that after the destruction of the one church to Saint Anne in the 7th Century, there was none in her name in the peninsula for a thousand years. But the basic impression seems to be right. There were too many men around there in the patron-saint business; even if she was a native girl, she was forgotten and gave up, for a good long time.

The Morbihan is misleading in this respect, since its only famous pardon is at Sainte-Anne-d'Auray and a great deal is made of it. But the story behind that is not altogether satisfying. You can't say the department is poor in pardons, it has hundreds every May to October, bagpipes and all, for the pardon is as common to all Brittany as crêpes and crucifixes, and granite. But with the one exception these are homey little affairs, not the big "picturesque" attractions and perhaps sometimes almost the matters of real gaiety and pilgrimage they once were, that go on in Breton Cornwall, la Cornouaille, Tristram's and Saint Ronan's country. What happened . . . But it is sad to tell. The whole business at Auray is not only a recent, that is 17th Century, Counter-Reformation concoction, but a most uncharacteristic one, with a strong Italian cast to it that would go very well with the Mediterranean but is jarring in this hazier male atmosphere. It worked, it became exceedingly popular, but this isn't the Saint Anne of the 9th Century and the church, rebuilt in the 19th, is ugly.

What is said to have happened is that a poor and pious peasant named Nicolazic, living in the 17th Century outside . . . Well, you could make it up; besides it's all in the books, positively a literary favorite. Poor man with visions, skeptical clergy, persistence of visions, capitulation of clergy, re-establishment of Saint Anne. There's just one nice note. After saint shows poor man where to dig for her statue, and he has been finding the heaps of money she leaves for him to pay for her church, the people around begin hearing a noise as of thousands of pilgrims coming to the place although there is nobody to be seen, and so the word spreads that those are the swarms of the dead, coming to pay homage. That's more like it; that belongs. A somewhat similar story, although of quite different import, also in all the books, also origin of a pardon, honors Mary and features a village simpleton and tree-dweller named Salaun ar Foll, the Crazy, whose constant and almost only words are "Oh, Maria! Oh, Maria!" Vision; and after death the words are found written in a huge miraculous lily that grows from the corpse's mouth.

This is in the 14th Century, at the time when the soldiers of Blois and Montfort were ravaging the countryside and killing any wandering peasant who wouldn't declare himself. One such band is said to have addressed the usual "Qui vive?" to Salaun, who replied with unprecedented lucidity, "I am neither Blois nor Montfort, I am a follower of Lady Mary," so they laughed and let him alone. In other words, he was the perfectly rational type of the Conscientious Objector, in the only guise feasible at the moment. Hearing of the miracle of the lily, Jean de Montfort, no doubt sensing in it a deep current of the time, vowed to build a church on the spot if he triumphed over Charles de Blois, and immediately after the Battle of Auray (September, 1364) he did lay the first stone of the church later finished by his son. It is called Notre-Dame du Folgoët—Our Lady of the Nitwit of the

Woods. Incidentally, Charles de Blois turned out to have had certain saintly attributes, very troublesome after his death to the conqueror. In one version Jean de Montfort walks barefoot to worship at his fallen rival's grave; in another he flees in embarrassment at the sight of blood flowing from a wall, at the location of the heart in a portrait of Charles which he had had whitewashed.

In the Salaun story you feel a genuine need of Mary, of a principle of love and concord to set against the endless fighting and dissension, and perhaps also to set against the fierce figure of Jeanne la Flamme, in whom motherhood was all warrior pride. La Villemarqué didn't make her up. She did burn the French camp at Hennebont and very likely did exult over the human cinders, as the ballad has it, in anything but Christian style. It would have been in character.

At that time, time of Jeanne and of Salaun the Crazy, the man of peace with his vision of Christian love, La Villemarqué relates that among Bretons of both Wales and Armorique on certain solemn occasions this ancient ritual cry was still to be heard: "No, King Arthur is not dead!" He was the incarnation of autonomy, and in Brittany it was said that when battle was imminent the armies of Arthur could be seen crossing the crest of the Montagnes Noires. They must have been appearing there most of the time in the 14th Century and probably didn't stir up the frenzy of enthusiasm associated with the sight in the 7th—that is if the date given for the following, from "The March of Arthur," were to be trusted. It isn't, and this is one of La Villemarqué's most suspect as well as splendid productions; however it conveys a thoroughly plausible atmosphere, and Arthur is after all very deeply a Breton property. The eye-for-an-eye notion looks quite decadent in comparison.

> Heart for eye! hand for arm! and death for wound, in the valley as on the mountain! and father for mother, and mother for daughter!
>
> Stallion for mare, and mule for ass! leader for soldier, and man for child! blood for tears, and flames for sweat!
>
> And three for one, that's what we need, in the valley as on the mountain, day and night if possible, until the valleys are flooded in blood.

The same unmitigated savagery appears in another, equally suspect song attributed to the 6th Century, celebrating the Breton foraging expeditions for wine into Frankish territory, and which La Villemarqué claimed to have heard thirteen hundred years later as a tavern drinking-song. No doubt he did in part, or something like it; perhaps it was dregs he found and bewitched back into wine; the creation anyway has its own veracity, possibly of a higher order than the scrupulous "folklorist" could give. So let's say for the pleasure of it that we are effectively in the 6th Century, as hypnotized out of the 19th.

The power of the bards was slipping at the time, and there were various cleavages in their tradition, pagan-Christian for one, but the form here is pure bardic in both parts, the song being described as a welding of two. The second is supposed to be a war hymn in honor of the sun, taken from the sword-dance ritual that Bretons shared with the Gaelic and Germanic peoples. You have to see it, in fact the words force you to, though more in the wild alliterative three-beat original—the warrior throwing his sword in the air and catching it with one hand while with the other he tosses back his long flowing hair, and in the other scene the whole bloody drunken crowd reeling home with their loot. Such were the men those first saints, arriving in their little boats, stone or otherwise, from across the Channel, the first of them in the long beautiful bay of Saint-Brieuc which remains haunted by them forever, had to deal with. No, it was not a job for a lady, nor for a gentle type of man either, even aside from the wolves, and aside from the dwarves, korrigans and so forth. Especially as what you hear in this is not just the yawp of the brute; mystical powers and sanctions are in its throb; it is possible to feel in peril oneself under the fierce naming of objects and elements, which seems to hurl one back into the very origins of poetry.

Chronologically, at least to keep up the grand illusion, this is some nine hundred years *after* Aeschylus and Euripides, and twelve hundred or so after Homer.

Better white wine of grape than of blackberry; better white wine of grape.

Oh fire! oh fire! oh steel! oh steel! oh fire! oh fire! oh

steel and fire! oh oak! oh earth! oh waves! oh waves! oh earth! oh earth and oak!

Red blood and white wine, a river! red blood and white wine!

Oh fire! oh fire! etc.

Better white wine than beer; better white wine.

Oh fire! oh fire! etc.

Better sparkling wine than mead; better sparkling wine.

Oh fire! oh fire! etc.

Better wine of the Gauls than of apples; better wine of the Gauls.

Oh fire! oh fire! etc.

Gaul, vine-stock and leaf to you, oh dung! Gaul, vine-stock and leaf to you!

Oh fire! oh fire! etc.

White wine to you, Breton of stout heart! White wine, to you, Breton!

Oh fire! oh fire! etc.

Wine and blood flow mixed; wine and blood flow.

Oh fire! oh fire! etc.

White wine and red blood, and fat blood; white wine and red blood.

Oh fire! oh fire! etc.

It's the blood of the Gauls that flows; the blood of the Gauls.

Oh fire! oh fire! etc.

I have drunk blood and wine in the rough fight; I have drunk blood and wine.

Oh fire! oh fire! etc.

Wine and blood nourish the drinker; wine and blood nourish.

Oh fire! oh fire! etc.

II

Blood and wine and dance, to you, Sun! blood and wine and dance.

Oh fire! oh fire! etc.

And dance and song, song and battle! and dance and song.

Oh fire! oh fire! etc.

Sword dance, in a circle, dance of the sword.

Oh fire! oh fire! etc.

Song of the blue sword that loves murder; song of the blue sword.

Oh fire! oh fire! etc.

Battle where the wild sword is King; battle of the wild sword.

Oh fire! oh fire! etc.

Oh sword! oh great King of the battlefield! oh sword! oh great King!

Oh fire! oh fire! etc.

May the rainbow shine on your face; may the rainbow shine!

Oh fire! oh fire! oh steel! oh steel! oh fire! oh fire! oh steel and fire! oh oak! oh earth! oh waves! oh waves! oh earth! oh earth and oak!

The last of the Roman legions had pulled out about a century earlier, and much relieved the poor fellows must have been by that time, to let the saints take over. Among other angles, the province was spooky; the islands of the dead were right there off its westernmost tip but a lot of the departed didn't seem to have gotten there yet, they were all around, if you took so much as a stick from a wrecked ship they would drag you down to one of their places under the sea. And such manners, such songs.

But the clash and bang are lost in our language. The refrain of the above goes like this in Breton:

Tan! tan! dir! oh! dir! tan! tan dir ha tan!
Tann! Tann! tir! ha tonn! tann! tir ha tir ha tann!

It might be true that Bretons are unusually devoted to Mary, having been so unusually deprived of her, by history and cast of mind.

THE cast of mind is westward, to land's end and the sun sinking into the sea. To such geography the sun, whether a god or not, is godly, and the feminine principle turns up mainly as trouble-making, that is in the figure of the witch. This might apply to California too, only it's not Celtic and it lacks saints. It is hard to be a first-class witch without a saint nearby to keep you stimulated.

What stature and stride they have, those 5th and 6th Century saints, driven over from the British Isles by the Angle and Saxon conquests, to the dark enchantment-ridden peninsula where a holy man must know first of all how to domesticate a wolf. A common knack; Saint Hervé, the blind one, was only one of many who had it. They are Christian, of course, but it is not until centuries later, up to the time of the lawyer Saint Yves, that their miracles begin to be of the general Christian character. In the beginning they move in a fairy-tale world, which we know well from our dreams and the brothers Grimm, and where such Christian traits as turning the other cheek are not in order at all. If people are not nice to them they put a curse on the land so no tree will ever grow there, or take some other revenge; there is hardly a story of forgiveness in the lot, in either of the two categories of saints. There were the doers and organizers, the political type, and the anchorites. The first generally set up a community, the *lann*, around a monastery, in distinction from the *plou*, the non-religious community built up around one of the immigrant war leaders or tribal chiefs; hence the prevalence of the two syllables and their variants (*plo*, *ple*, *pley*, *lam*, etc.) in Breton place names. But doers and brooders alike must reckon with sorcery. The enchanted forest Broceliande where the genuine original gold-digger, Vivian, did poor Merlin slowly out of all his craft, was a reality. As a matter of fact the forest is still there now, not quite in the same sense, in the inland department of Ille-et-Vilaine, which is not where one imagines it.

One thinks of it as nearer the sea, probably because Merlin, until the girl had him completely bound and bereft, wandered back and forth between there and England with such ease, and it was not because he was a magician. King Arthur, who might conceivably have died in Brittany as the patriotic version goes, seems to have moved armies back and forth just as easily, as if it were over the next hill, with no water to cross. "Crossing the sea," Saint Gildas is said to have said, "crossing continents, doesn't frighten Bretons; on the contrary, they like it." But really it is more that they *step* across the Channel, and when the saints go to Rome, as Gildas and Cado did, they seem to do that in a few steps.

Gildas, whose *lann* as we were saying is only an air mile or so from Locmariaquer, is in the sociable or man of action category. Saint Ronan, whose church and traditions are up beyond Quimper, is one of the most appealing of the anchorites, especially useful against stomach-ache and spells cast by tailors. The witch in his story, named Kében, was not very witchy, in fact only an ordinary virago, but troublesome enough. What riled her was the saint's influence over her husband, who should have been doing his proper peasant work and not contemplating the beauty of God's world, for Ronan was not even the helpful sort of hermit who could make springs gush forth and harvests grow in a waste field; he was the pure meditator, wanting only solitude. So Kében hid her youngest child in a kneading trough and put it out that Ronan, in the shape of a wolf, had run off with the little girl. People believed her. The saint with his bell, a usual prop with Celtic monks, had caused annoyance at various pagan gatherings; also the stone boat he had arrived in was vaguely in the shape of a mare, and this stone horse with which he is often portrayed had a way of leaping around the hillsides, showing him where to go. He was a sorcerer, clearly. A trial ensued before King Gradlon, who was having a bad time anyway with his wicked daughter and the City of Is, shortly to sink beneath the sea. He behaved badly in this case, set his dogs on the saint and was much embarrassed at their falling humbly at his feet. Ronan told the people where to look for the lost child, who had meanwhile suffocated; he revived her, and was thereafter so sought out by the king and everybody else that he had to flee with his stone horse all the way across the peninsula, to have peace again. He died over there, in the still resonant village of Hillion across the bay from Saint-Brieuc, and remained so stubborn even in death, it was left to the oxen pulling the coffin to decide where he wished to be buried.

There was no flying or leaping that time. Slowly, for days and days, the beasts made the long trek back to what we know as the big tourist attraction in his name, Locronan. He had asked forgiveness for Kében the other time, one of the rare instances of the kind, but now when she bashed off one of the oxen's horns and climbed up to spit on his dead face amid shrieks of abuse, his patience gave out, the Celt overcame the Christian and he caused her to disappear into the earth.

Her contemporary Dahut, King Gradlon's daughter, was an equally familiar type of witch, the classic nympho, good-looking, sex-crazy and sick in the head, and she too probably wouldn't amount to much if there hadn't been a saint in the picture. That was the great Guénolé, or Gwennolé, advisor to the king and founder of the first monastery in Brittany. In certain versions Dahut is a real witch, or a fairy who turns into a witch as so often happens, and not the king's daughter but his lady-love. However she became standardized as his daughter. After one of her nightly binges she steals from her sleeping father the golden key of the sluice-gate protecting the City of Is, lets the tide in, just for kicks evidently, and is drowned when Saint Guénolé orders the fleeing and unhappy king to push her off the back of his saddle. Or she can be made to look more rational, as a last high priestess of paganism, driven to suicide after the king's conversion and deliberately taking the city with her. She became Ahès the plaintive mermaid, the Breton Morgan le Fay crying across the waves, and nobody knows where the city, Chris or Ker-Is, is, some say off Douarnenez, although in Wales and Ireland the same story may be associated with a lake.

Lake or sea, the point is not wickedness, not Sodom and Gomorrha, so much as water, for the same reasons of the soul that have given the springs of Brittany their fabulous connotations. Water for memory—a nebulous sense of metempsychosis, a lingering on of presences, a living with realities beyond the visible. "It was I who gave Moses the strength to cross the River Jordan [sic]; I saw Sodom and Gomorrha destroyed. I have been the standard-bearer of Alexander. I know the names of the stars from sunset to dawn . . ." This is attributed to the great Welsh bard and prophet Taliesin, one of those said to have emigrated with such chiefs as Gradlon himself, and to have been converted later by Saint Gildas. All that is probably a fabrication, but the words are in his spirit, and his influence was undoubtedly felt in Brittany through other emigrant bards even if he never got there himself. He would have been speaking in earnest, and the people, although not claiming so much for themselves, would understand. The past is with us, the journey of the dead will not be done until the end of the world, and therefore wickedness is rarely beyond redemption and tends to wash out in a pathos common to all. Otherwise how could Dahut become Ahès, the anadyomene, the weeping daughter of the sea? She is still evil, when she is seen it means that a terrible storm is coming, but you can fear her without hating her, as if she were victim more than agent. It appears too that when the bells of the sunken city are heard, a mass is being said there, and if a passing sailor, or somebody, were to go down then and speak the final missing words of the mass, the inhabitants of Is would all be freed at last.

There is another city of the kind on the north coast, imbedded in what is known as the Great Rock, off Lieue-de-Grève. Every seven years at Christmas time the top of the rock opens at night, showing brightly lit streets below, and if anyone were bold and athletic enough to go down into them on the first stroke of midnight and be back out on the twelfth, the city would be brought to life again.

As for the end of the world, it has been dated with surprising coincidence by two different means of calculation. On a little island off the north coast there are two granite crosses some distance apart, which are said to be drawing together by the length of a grain of wheat every seven years, and the world will end when they meet. The other prophetic object is a single huge cross on a beach, not far from the mysterious city in the rock, which is sinking by the length of a grain of wheat every hundred years and will finally vanish under the sand to mark the same event. Either reckoning is said to give us between 300,000 and 350,000 years to go.

But getting back to Gildas, who belongs more to our story, being so close a neighbor, magic surrounds him too but differently. He is the priest politician and diplomat, not so much bothered by witches and dragons, perhaps even rather obtuse in that sphere, but grand in his own line of work. The full story of that jump of his is that he got the devil to swallow a big darning needle at the end of a long string, then jumped and dropped him in the sea. This was the work of a strategist, as distinguished from the saintly exorciser pure and simple. Another antagonist was the actual if devilish tyrant Conomor, who comes down in legend as the Breton Bluebeard, really a clan leader who started with a reasonable idea of unifying

Brittany and became crazed by territorial ambition. Gildas, to his sorrow, helped to arrange a marriage of power for him with a girl named Triphine, daughter of the duke of the Vannetais, the region between Vannes and Nantes. That much seems really to have occurred, and so did the Council of Ménez-Bré later on, in the year 548, by which Gildas got Conomor excommunicated and deposed. In between, the story is the perfect blend of the three elements of this spiritual landscape, history, fairy tale and Christian miracle, including: ring given Triphine by the saint, which will turn black when danger threatens; five previously assassinated wives of the tyrant, giving advice from their coffins in the cellar; resuscitation by saint of murdered Triphine—the least he could do after making such an awful mistake, though she dies again shortly, after giving birth to her son; decapitation by tyrant seven years later of little son Trémeur, who picks up his head and is able to walk with it as far as his mother's grave, before expiring. Both the child and his mother became saints, of the category of martyred innocents, a nearly total contrast to the great operators such as Gildas or centuries later Saint Yves, but at least as powerful in the long run. In ballads, plays, architecture, the "Mystery of Saint Triphine" has gone on reverberating almost until now.

And right along with it, to the same ears, through so many days and nights and generations, there have also been echoing from time to time the bells of the City of Is, out of "the dolphin-torn, the gong-tormented sea," only that's from another Celt, another piece of sea; around Brittany there are not many gongs or dolphins. But there are other sounds and meanings, as Debussy knew, very difficult for our cluttered minds and essential to the oyster business. The importance of bells in general has to be grasped. The mind must hear them; it ranks with the capacity to déguster, for understanding the subject. In the ballads the Breton soldier or sailor or crusader far from home and perhaps dying, or the girl about to be raped by an Englishman, mourns for "our fields, our saints, our bells . . ."

Gildas not only carried a bell like all his colleagues but was a skillful forger of bells and rather vain about it. One of his productions was a source of friction with his friend Saint Cado, but that must have been a curious friendship anyway, for aside from fooling the devil the two had little in common. Cado, later martyred by barbarians in Italy, was a far more gentle and sensitive type. Gildas once came on him reading Virgil and weeping that so beautiful a spirit should have had to be damned; Gildas threw the book away and said in effect that the poet had been nothing but a pagan and to hell with him. Of course the same story is told about innumerable pairs of eminent Christians; it is a way of defining an eternal difference. Well, Gildas stopped by one day with a bell he was particularly proud of, and refused his friend's pleas to give it to him, saying he was taking it to Rome to put on the altar of Saint Peter. But when he was showing it to the Pope it went dumb, not a sound would come from it, and hearing who had bewitched it the Pope made him take it back, with his blessing, to Cado, an even more frequent pilgrim to the Holy See. Under a different attribution, it is said to be the one that is kept in a little church in the Morbihan, to be rung in the ears of deaf people.

A more generally useful bell in the Morbihan, a great big strident one probably lacking in the subtler metals but with a good bass, hangs in a little open pavilion above the pretty chapel of Sainte-Barbe, overlooking miles of lovely hills and with no road to it, only a long worn track through fields. Then a grove where you stand at steeple height, the chapel clinging insect-like to the steep shoulder of hill below. This is in the region of Le Faouet and only a few miles away are two of the small-scale architectural joys of the world, with sculpture in them to match, of such wry cranky genius and in settings of such delight because natural and beautiful and absolutely unannounced, one a farmyard, the other a monochrome village going about its business and not asking to be noticed at all, you can never forget them or remember them without smiling. They are the little churches of Saint-Fiacre and Kernascléden. Sainte-Barbe is not in that league as architecture, and has no sculpture, therefore no comedy, the Breton gift in stone having always a gnomic and funny streak when it is most glorious. This place is just awfully pleasant, it makes you feel good, and there is the bell; you pull on the rope, and on a Sunday in good weather it will be ringing all the time, bang bang bang over the wavy fields. No sea; this is inland.

It is for luck, in some obscure way, like drinking at the sacred spring there. What they say is that the bell wards off storms. But it must do more. One of the main purposes of the saints' bells was to ward off korrigans.

THERE is one more witch who has to be mentioned, and a priest who was no saint at all. Their joint image is very close, just across the strait like the huge striding image of Gildas and in fact in his very place. It shows what the Breton mind can do when really pushed, for she under other lights is the girl of your dreams, and he the great star-crossed lover, Abelard.

Poor fellow, he does suffer from appearing that way, as in a double exposure, against such a figure as Gildas. It is a cruel irony that he should have had to serve out his ten-year exile as abbot of that particular monastery, as if his disgrace and castration and other troubles in Paris hadn't been enough, and even if the saint for whom the place was named had been dead some six hundred years, so at the time there was probably no very lively comparison to be made. It is now, after another seven hundred, that one tends to see them there side by side, so invidiously. However, even if anybody had looked at it that way, it could hardly have made things more unpleasant for Abelard than they were already. As a matter of fact even the love story, the perfect model of the grand passion in our usual versions, seems to have had a very seamy side. Abelard was undoubtedly brighter than his great enemy Bernard of Clairvaux, and in his early days there must have been something attractive about him, beyond the blazing rationalist intellect that brought him as many partisans as enemies. It is virtue that is missing. He was vain, self-indulgent, as bad in his way as King Gradlon's daughter. There is some evidence that after marrying Heloise, which he had no business doing as she told him herself, he forced her into a convent not out of guilt but because he was tired of her, and went back to chasing other women in Paris. So it would have been his infidelity to Heloise, rather than his seduction of her in the first place, that impelled her uncle to take his terrible revenge.

In any case, the mutilation did not improve Abelard's character; it only brought a stop to his chief amusement, and if anything made him a little sicker in the head. In his account of his own misfortunes, evidently written during his years at Saint-Gildas-de-Rhuys, he shows himself up as paranoiac and generally odious, so that one rather sympathizes with the monks who wanted to poison him, even if everything he wrote about them was true. He said they were thieves, assassins, drunkards, filling the abbey with their children. That last touch makes one wonder; it makes them sound like unusually solicitous fathers, if all the rest were true. What he might have added is that they were Bretons, Abelard himself being one of the rare cases in Breton history of the total renegade. He came from down around Nantes, a section that had remained predominantly Gallic, in contrast to what is called "la Basse Bretagne," Lower meaning western, not southern Brittany, but he had no use for his own region either, except as a place to hide Heloise when she was pregnant. The Breton language was abhorrent to him, he couldn't speak it, called it barbaric—the same adjective used now by André Malraux for Breton sculpture. That was Abelard's view of everything Breton, and considering the saints, the bells, everything his blind arrogance rejected in that place . . .

The monks may have contributed something to one ultimate piece of revenge. Under the lash of his contempt they would certainly have done some talking in the neighborhood, and the field was fertile. Aside from witches and evil spirits in general, there had been at one time a college of druid priestesses on an island off the mouth of the Loire, whose traces were far from obliterated in the 12th Century, and were showing up in a prevalence of sorceresses around the place even two hundred years after that. That was right by Abelard's home ground, where Heloise went to have the baby. The druidesses were supposed to have had supernatural powers very like those claimed by Taliesin; they could assume the forms of animals, foresee the future, create storms at sea, cure fatal sicknesses. Furthermore book-learning in ladies was suspect to begin with, even without scandal; it smacked of unholy powers.

So what with one thing and another, including some hurt Breton pride, Heloise turns up as one of the ghastliest witches ever to appear on earth. She can read and write, speak French and Latin, say mass and also prevent the priest from saying it, find gold in ashes and silver in the sand.

I change myself into a black bitch, or a crow, when I want, or a will o' the wisp, or a dragon;

I know a song that will split the heavens and make the ocean shake and the earth tremble.

I know, myself, everything there is to know in this world, everything that has been and will be.

The first brew I made with my sweet clerk was made with the left eye of a crow and the heart of a toad;

And the grain of the green fern, picked a hundred fathoms down at the bottom of the well, and with the root of the Golden Grass plucked in the field,

Plucked bareheaded, at sunrise, barefoot and in a nightgown.

The first time I tried my brews was in the ryefield of the lord Abbot:

From eighteen bushels of rye the Abbot had sown, he got two handfuls for his harvest.

I have a silver chest at home, in my father's house: anyone who opened it would be sorry indeed!

There are three vipers there, brooding on a dragon's egg; if my dragon hatches, there will be havoc.

If my dragon hatches there will be great havoc; he will throw flames for seven leagues around.

It is not partridge or woodcock flesh, but the holy blood of the Innocents, that I feed my vipers.

The first I killed was in the cemetery, about to be baptized, and the priest there in his surplice.

When they had taken it to the crossroad, I took off my shoes and went to unbury it, silently, in stocking feet.

If I remain on earth, and my Light with me; if we stay on earth another year or two;

Another two or three years, my sweet friend and I, we will turn this whole world inside out.

—Take care, young Loïza, take care of your soul; if this world is yours, the other belongs to God.—

La Villemarqué had a hand in this, but it is one of the cases where he seems to have had a good deal to start from. According to him, it was still one of the most popular of all Breton songs in his time; it was sung all over Brittany, but had originated in the dialect of Vannes. Sic transit grand romance, in that climate. Not that there aren't plenty of touching ballads of love, but the lovers are supposed to be virtuous.

THE bell of the little chapel of Saint-Pierre rings harsh and jubilant across the sweet flowering fields and the manure-fragrant habitations, Ker-this and Ker-that, and the white stacks of oyster tiles nearly completed by the shore, one finger of land away. It might even penetrate, though dimly, to the main necklace of the same white piles on the riverside, over beyond the bourg. A joyous sound, because heard only this once in the year and it happens the sun has come out for it in spite of a brisk and chilly breeze that sets up a giggly flutter among the nuns, all but the Mother Superior, as though they were being turned into trees because of their pretty faces. The regular town church-bell is ever so much deeper.

For a couple of days people have been cleaning and airing the chapel, always dank after the winter. There are flowers on the plain little altar, and on the beach a few hundred yards away the pyre is ready. The chapel has a ruinous look because the big stone buildings to either side of it have been unused and falling apart for the last century or two, and nobody has time to cut the grass and weeds around it, but it is intact, such as it is—modest, unbeautified, interior ornaments few and cheap. It has neither the fine beamed ceiling of most Breton chapels, nor any speck of sculpture of the great period, 16th Century to early 17th, when the province had its epidemic of genius in that line, nor any other distinction, except the inescapable ones of texture and rightness of line. The two extant farmhouses in Saint-Pierre, belonging to the Giannots and the Janecs, Yvette's parents, look huge as seminaries by comparison, only the Janec thatch began to leak last winter after some two hundred years, so the old man, with help now and then from a son or son-in-law, is replacing it with slate, and at the moment the windowless attic part is exposed the whole length, showing how tiny the human living quarters are. The essential point is that the structure is granite, and although for the same pains and expense the old man could probably build a new little stucco house, or he could sell out to summer people and move away, he is not going to.

The chapel begins to fill, just as the Janec monkey, brought home by a sailor son-in-law, escapes to the yawning roof rafters with a soup bowl. The children howl with delight, while the bell continues to ring, exciting the monkey further, but the Abbé starts mass just the same and soon order is restored. The curé, of course, was supposed to officiate, but is ailing, to nobody's regret. Marie-Jeanne is one of many who have come from the bourg for the pardon, not as many as used to before the last war but still too many to get into the chapel. She is in the cluster outside, following the mass through the open doors, beside Mme. Aurogné who wears the uniform black dress and sweater and sober coiffe of the older women—a rarity for her, what with oyster-work and the farming the rest of the time. It is the women of the bourg, the storekeepers, who dress like that every day. Up around Concarneau it would all be very gaudy, with bright colors and velvets and starched white like sugar palaces, and the same white would be seen any day in the week on many bent old figures chopping in the fields; not here; fancy never ran to costume in the Morbihan. Another point of tourist interest is lacking too. There are none of the gnarled, violent, Romanesque faces that you see in some of the fishing ports or among the farmers of the Côtes-du-Nord. The mustachioed old boozer Janec, Yvette's father, who can't be bothered to go to a pardon in his own front yard and is over there rewarding himself for his week's labors with a bottle of applejack right now, comes the nearest to it, but is really not of that explosive stuff at all. In the gathering at the chapel it would be hard to tell piety from comfortableness; perhaps they come to the same thing; the service is a natural fact and so is the nice big comfortable girth of a woman over forty, nothing to be in a twit about in either case.

The air of festivity, which is real, is a formal and quiet one. People are looking their best, but you feel it was done easily, and no woman there has tried to disguise her age, or hide her hands. The hands and feet are all big and the legs generally stumpy; that far the anatomy is common even to the sheltered ones, like Marie-Jeanne and the Postmistress, though an indoor thinness of skin and smoothness of finger joints sets them apart.

The Giannot girl's hands are tough and discolored as a man's from farm work; with her trim hair-do and spring suit and unbusy-looking smile she would do well on any boulevard, but it must be years since she has been able to take off her wedding ring and her high-heeled shoes have to be man-sized, from a work-life in sabots. Even so, hers aren't the hands of *tout le monde* here. The hands of the oyster-workers, men and women, are more or less chronically bruised, swollen, blotched—not evenly weathered, not evenly swollen either; there are usually chilblain scars and at least an incipience of arthritic knuckles. The deformity is mainly from cold salt water and handling the rough shells; the people are too skillful to cut themselves much with shell edges or the sharp tools of the trade. Only if they cut a finger at home, as Yvette does now and then from her husband making some rowdy drunken scene when she is cooking, it takes a long time to heal, at least in winter when the chantiers are bustling, and leaves another welt.

So there they hang now quietly, pair after pair of these raw bumpy hands that can seem for a minute strangely on their own, speechless and with a separate character like animals. Nails are ground to the nub; a little finger almost resembles a thumb; you wouldn't believe the same fingers could be as quick and knowledgeable as a musician's in February and March, when it comes to removing the tiny fragile oysters from the tiles without hurting them. Perhaps the hands have a pride that the head scarcely knows. Anyway the heads taken all together have brought a massive dignity to the little fête, that is like a reprimand to those poor dumb appendages with their story of pain and hardship; you there, be quiet, this is not for you. Or is it for them, for them too, as it is also for Jean-Pierre Aurogné the almost idiot, who like them can merely be scrubbed for such an occasion, having no other possible dignity to call on, and whose hands are bandaged because of what happened at the hedge Thursday? He stands back from his mother, among the men, and she approves of that and is proud of him for it, although he shows a little strain of conflict between the Abbé's strong nasal chanting from inside the chapel and the similar whine of the coast guard plane passing over. It passes, and the Abbé's voice comes out with sudden new vibrancy, on the verge between true music and some wilder form of statement, and the full-throated responses have the same ambiguity. It is a sound of good loud human singing, sure in pitch and unequivocal in spirit, laced with the howl of a dog at a death in the house or ghosts of drowned sailors on a night of storm. Yet all is regular and proper; the ritual is not a form of self-expression but instead of it, so there is no threat to decorum, unless Jean-Pierre were to wander away or the monkey act up again.

The funeral, and of a man so young, was only day before yesterday, and in some degree everyone there has suffered loss in it. Like or dislike, gossip or weep, it doesn't make much difference, the organism is hurt by a wound to any part and it is as though the idiot in his twin-brother delusion, grabbing the fistful of thorns, had acted for everyone. He acted moreover, perhaps under some influence emanating from the Abbé and equally unrecognized by the two of them, out of the brotherhood most particular to this place, that of the hands. The hands of Locmariaquer, all thumbs as they may seem but not as we mean it, and green as God's sweet springtime in the science of flowers, have bled through him for the young man who is so absurdly dead; whose hands until Wednesday night shared their grey anonymous honor. Now the bell calling everyone together here, and the fields sprinkled with poppies, and the sea itself, have become their voice.

The sea laps at the stony bank a little way toward town, a yard or two from where Françoise is seated on the ground beside her mother. Later she will limp over to Saint-Pierre as she does every evening for milk. She can't do it twice and hasn't clothes for a fête anyway, not even a coiffe any more; what she wears, today and any day, is the little white cap the coiffe would be pinned to if she still had one. She doesn't go to church in the bourg either, it is much too far; the Saint-Pierre people take care of getting bread for her every day. But the bell brings her the fête, as the bell of the bourg did the funeral, and she and her mother are soothed by the human confluence, belonging to it even if they are not there. People will come by and chat later, perhaps the nuns, or others; what she is mainly hoping is that somebody will take the last of the puppies, else she will have to kill it before it starves and she can't bear that.

Another woman who can hear the bell, or at least is in earshot of it, is young Mme. H., wife of the biggest and youngest of the *gros richards*, over at the point the other way, where the great stacks of tiles are. Juicy, kittenish, manicured, spoiled, dressed in flawless white sharkskin slacks and at home only in her Fiat sportscar at a hundred and thirty kilometers an hour, to her what is unbearable is the presence of her two young children. She can stand them for an hour but today has had them for six and is not only ready to scream but screaming, at them; hugs, candies, threats have failed as usual. Luckily she doesn't have to suffer them often. Her mother, whether from love of the children or respect for the oyster business, keeps them in Auray. The H. business, with a main center at Riec-sur-Belon and two chantiers on the north shore besides the one in Locmariaquer, does command respect; so does Monsieur H., aged twenty-nine and the most enterprising oysterman of the region, the only one who has mechanized on a large scale. His mother helps to run the ancestral establishment at Riec, which no doubt impels his wife to have a finger in the pie too, although she lacks the deep commitment of her husband and mother-in-law and secretly would be happy to chuck oysters altogether if they could have the same station in life without. Her husband, she feels, is tyrannized by them; oysters, oysters is all she hears; he has slides, reads pamphlets, goes to conferences, experiments all the time, with some new wood for stakes, new kinds of collectors, new machinery; she has been wheedling at him for five years to take her on a vacation to Argentina but the oysters always manage to prevent it, even in summer. For a handsome and sociable young man, even fun-loving in a way, it is crazy; he will never take a vacation at all.

However the compensations are great and so are her powers. That she should wrest him from the dignity not to say wizardry of his calling is unthinkable, but between sex and her maniacal skill at the wheel she has him pretty well enthralled. His life has to be largely on the road, crisscrossing the peninsula between the various chantiers, and convinced of needing her beside him, if not of her usefulness in the business, he puts down what regrets or embarrassment he may feel about his children. Not for her the life of the sailor's wife endlessly waiting, commonest picture of the women of Brittany, though she is Breton too. She is a poor little Vivian, a cheap replica, with a shimmering convertible for her palace and her Rh factor on a medallion around her neck—the one precaution she would agree to—and for her song the scream of the brake. It sounds faintly now across the fields as she rips through the nearest hamlet, taking her unmanageable brats back to their grandmother. It might be, since she was brought up piously and knows the Saint-Pierre families very well, that the little pardon is making her extra irritable, as it is old Mme. Hémon at her window above the store in town. She could perfectly well have gone but prefers to dwell on her resentments at home, greenly following from afar every cadence of the mass and her daughter's guiltless devotion.

The chapel empties; the procession forms; breeze again, through pale flurry of sun-gauze and twinkle of poplar leaves, but the nuns are in sober mood, herding their thirty or forty little girls into a line behind the cross. Then slowly wafted on oxen tread the corporate body lumbers chanting over the hard-packed stony track to the beach, calling in shrill quarter-note modulation, almost Moorish, on its various saints for various blessings. "Bless our hearths and our children . . ." "Bless our fields and our animals . . ." "Bless our boats and our harvests"—harvests from the sea; first the Abbé's voice alone, not resembling the voice of teacher, friend, athletic director but piercing, remote, voice of a man no longer a man but invested lungs and all with the ancient authority of the intercessor; and over and over from the others the same refrain ending in the same down-slid note, the bleat of the human question mark; until all are grouped on the sand and after a final poke or two from the pyre-wardens the Abbé applies the match. Smiles here and there at the recollection of the curé setting his robes on fire last year; hubbub and shrieks of laughter from the young, to frown and cluck of Mother Superior, as the smoke bucks straight into the crowd; order once more and the last and loudest song of prayer, to make it rise over the thunder of the flames and be audible not just there, somewhere else, above, below, beyond, down to the poor spirits of the drowned clinging to the carcass of their ship along the beach, out to sea, out to the ears of the living from almost every family who are riding on the seas, and perhaps to the western island of all the dead, *les trépassés*, those who have gone beyond, last Wednesday night in a ditch or any time.

Why the fire? Nobody remembers; it is something that has always been done, and whatever spirits are meant to be placated, it is a rather jolly way of doing it, though reverent. Everyone is quite gay and chatty afterwards, as though a load had been lifted or a communal statement made to everyone's satisfaction. Françoise too over by her own beach, reporting to the ear of the old black-shawled hulk beside her that the smoke has gone down behind the trees, makes it sound like comfort, and the mother, all bulk with just the one spider-strand of life left in it, nods in approval.

The missing words: the question unasked, the last phrase of the mass not supplied, whereby the curse would have been lifted and all saved. The curse is not on the seeker, the duty he can't face is for the public good, not his own, but how can he live with his failure afterwards when so much depended on him? There must be some will in him to have his jaws locked and his tongue tied, so that the misery will embrace him too; there is no other way to explain it. This is our problem, the burden of the missing words, which have never yet been said.

The mysterious castle is always in the woods. The young man who has lost his way is well treated but cannot understand what is expected of him. There is a king too ill to do anything but go fishing, and another invalid even older in another room, to whom a wafer is carried on a large platter called a grail, big enough to hold a boar's head. In the morning the castle has vanished and there is desolation. The young man obscurely knows that if he had asked what the trouble was, the king would have been cured and the waters would have flowed again through the dried-up land. But he had been taught good manners and that one does not ask personal questions. Another possibility is that at heart he was not interested.

Or it is the City of Is, under the sea. At that point in the liturgy there was dreadful silence and they were all looking at him, waiting for him to speak. He knew the words, he must have known what was needed and that the lovely city and all its people could have been made to rise from the ocean floor—by so little! but he could not speak. Their accusations ring in his ears to this day. How different from young nightmares, where it is only your own soul you fail to save, or the later ones of your wife or child slipping from your hands under the sea or into the fire. The true missing words are those of the quest, that would have redeemed a city or a sick land with all their people, without the hero—the knight, hitchhiker, sailor, dreamer, drunkard, man of God, worrying about himself.

"Canst tell how an oyster makes his shell?" The fool could ask, knowing the answer, the grim "No." But if Lear could have replied cheerfully, "Why yes, it's like this," and told him all about it, perhaps that old king too, like the fisher one, would have been hale and undeluded again, the good daughter would have been understood and the bad ones suddenly good, and the fresh streams would have flowed again through the land.

SEVEN

FOR SEVERAL WEEKS before the great day, when the movement of barges starts, the main work is sedentary. All along the shore the little groups of women, sitting on crates or anything else that will serve, are scraping and tying the tiles. The scene, for several miles either way from Locmariaquer, takes on a repose quite different from any other time of year, pictorially. There are tensions and murmurings as of beehives if you move close in; the hands and torsos are driven, and the faces half-hidden under a double protection of scarves and sunbonnets of newspaper, with only here and there a straw hat, are not in repose. The stacks of tiles that have to be dealt with before the time is out are appalling; it is hard to see how it will get done. But looked at from a little distance it comes close to a school of painting more Flemish than French, only the quality of light would prohibit the Flemish vision here, and the deep sense of flux too. The groups of women are stationary, except when someone gets up for a new supply of tiles or wire, but no forms are stable in this light; they tend to recede into the shifting ambience of clouds and the tide. It is colors that stand out. This explains the abnormal toyshop gaiety of the flowers, even aside from their variety and number, and the fantastic weight in the landscape of the stacks of tiles. By the Mediterranean they would look a lot smaller and would hurt your eyes, which would be a pity; it isn't often we get a chance to understand white.

They are nowhere near as white as they will be at the end, after the bouquets are dipped. There have to be some new ones every year and those are still russet; the whiteness at this stage, although extreme enough in the long view, is from last year's tiles, the ones not too broken to use again, and that are being scraped. They had one cleaning when the oyster seed was removed earlier in the year. Now the last scabs and blotches have to be taken off, with an instrument like a blunt knife, so the new chalk will stick. Children can do that part of the work and some of the older ones do on Thursdays; they earn one franc an hour, a third less than their mothers. The trick is getting the wire quickly through the holes, pair by pair until you have the twelve with the twist at the top, without fumbling around. It looks easy because the women have done it every spring since they were twelve or fourteen years old and some are in their fifties and sixties and still doing it, but it is not easy and the bouquets are not as light as they look. They weigh seven or eight pounds apiece even without the pickets and a good deal more later with the spat and muck on them; before it is through they will have been moved fourteen times, by hand, adding up to some two thousand tons per 100,000 tiles, the amount handled by a small yard. The big ones put out twice or even three times that many.

But the only time you get any sense of what this means is May and early June; it is the only time they are all on shore, and after the dipping, all white. The old year of the oyster is finished; the new one hasn't begun. Not that all the yards are on exactly the same schedule. Down from where Mme. Aurogné and five other women sit endlessly poking wire through the holes to form their bouquets, chatting all the while but never singing at their work—you never hear any casual singing across the water or the fields here, it is just for parties—a long line of people in rubber boots are trudging out at low tide along the right-angle paths to one of the parks; there was a delay there for some reason and they are still bringing in the last of the old naissain, way behind time. At a yard in the other direction, upriver, the bouquets are all finished and even limed, perhaps too soon; if the weather is either too dry or too damp before placing time, the coating may begin to chip or flake and the tiles will be spoiled.

The owner there is a middle-aged woman, who slowly paces among her stacks of tiles, keeping an eye on sky, tide and her one remaining worker for the time being, as though the dangers to the year's income were of a kind to be warded off by vigilance. Now and then she picks off a bit of dried lime between thumb and forefinger, crumbles it to test its consistency, appraises the sky again with much the glance she used to cast at her husband's mistresses. Her vigilance was equally futile there, his taste was so bizarre, as she learned way back in the beginning when she herself was only twenty and carrying her first child. Who could have imagined?—a little old cripple, ugly, simple-minded, forty if she was a day . . . She didn't mind the others so much but the name Françoise still stirs her to anger, although he has been dead two years and was the best of husbands in the basic respects; her son and daughter-in-law manage the restaurant with her now.

From where she stands, lacking only a pistol to be fully the picture of a military guard, and time has been when she nearly came to that after a run of thefts from the parks, she can see the aristocrat of the business, old Monsieur Hervé, in boots like hers and an ancient leather jacket, opening the sluice-gates under the millhouse to let the tide into his pond. An habitual twinge of envy, having nothing to do with oysters, pinches Mme. T.'s vitals at the sight. The inlet there, up to the stone millhouse and bridge which are of the 14th Century, was formerly one of the richest oyster beds in the area but is now so depleted, it is only by constantly regulating the currents from the pond that the old man can keep a crop there at all. However he is no longer ambitious and does well enough to supply his own private little Paris market, being one of the few who leave their oysters there to maturity. He can do it because he owns the millhouse and the pond beyond, nutritiously teeming and from which he can wash down his beds as required. But he has given up his tiles and buys the seed now; it was too much work for an old man, and his two sons were not interested. They are officers, at sea of course, one in the navy, the younger in the merchant marine; he has no reproach to make to them; they come home on leaves and are passionately attached to the place, only not to oysters.

Naturally he was disappointed but he took it in good grace, and when he thinks of some of the younger men in the business nowadays he sometimes feels his sons are well out of it. Young H. for instance, a likable fellow, admirable too in a way, who considers all the oystermen of the Rivière d'Auray more or less beneath contempt, chasing here and there all the time and with all that nearly losing his shirt last year in the Quiberon experiment. True, there might be something in that eventually. The idea was deep-water cultivation, American style, in the Bay of Quiberon, and the other two partners were still going on with it but in the section allotted to H. there was nothing but losses. And the life, the strain: for what? Monsieur Hervé cares greatly about the quality of his oysters, but is not concerned to have the biggest ones, nor the biggest business, neither of which would be possible in that location.

He has what suits him, as it did his father and grandfather, the latter having been one of the pioneers of the new ostreiculture and a friend of the great Monsieur Coste. And very pleasant it all is, the work and the life, except for the lack of grandchildren; so far neither son has seen fit to marry and Monsieur Hervé adores children. However there is still time, they will get around to it so his wife assures him, and meanwhile he always has a little friend or two from the neighboring farmers' or oyster-workers' families to go fishing for *écrevisses* with him in his "miraculous" pond, and who are not in the least awed by his noble lineage, any more than he is himself. The dearth of great castles in Brittany is symptomatic; the province is traditionally, through a variety of historical causes, the most democratic in France, the one where title and wealth count least. In this case the manor house, to call it something, is only shabbily lyrical in its setting of orchard in which cows graze, sloping toward the lift and fall of the river-mouth, also the neck of the Gulf for they meet there, with a secret-garden approach and a turreted structure at the end of the walled garden like a sleepy little joke at the expense of the true feudal wall—of Josselin, say, or Vitré—and the granite farm buildings attached at the rear, so that the large herd of Hervé cows are in effect under the same roof with the masters as in any farm around.

Inside the signs of gentility abound, but it takes a keen eye to pick them out, especially as the light is dim. There is an elegant little Empire salon, rarely entered except on cleaning day by the old maidservant clopping about in carpet slippers, who has been there thirty years and treats the French language like so many little bundles of grit and sawdust thrust into her mouth by hostile forces, or when Mme. Hervé, shy and lonely in her secretive attachment to culture, which has never had a chance to grow out of the girlish stage, creeps in more for a revival of memories than to practice, at her grand piano. The instrument is really just furniture; she gave up having it tuned years ago, it had to be done so often in that damp climate. The real life of the house, as in the homes of the poor, goes on around the huge old rustic dining-room table, on which a clutter of sewing, account books, etc. appears as soon as the meal is done. That is the real living room, with the TV (for this is the home of a long line of *gros richards*, of true oyster magnates of the old school, Monsieur's dress and kindly modesty notwithstanding) and the master's grimy old easychair which he will not allow his wife to recover, and the little leatherbound inherited library, of Dumas *fils* and such, under glass at one end, and the mistress's current reading matter, a curious compound of the pious and the avant-garde, the latter sent now and then by friends or cousins in Paris, scattered here and there to collect coffee and cake-crumb stains, for reading is actually more a vital notion than an occupation in her life and the books are usually half uncut when they get replaced by others after a few weeks; and for elegance, to come to the solid Breton heart of the matter, what emerges after a few minutes' conditioning and in total denial of the Frenchy little salon, which in this setting even more than the million others around the country where its exact like is found seems scarcely for humans at all, more a funeral parlor for Dresden china dolls, is the vast authentic old Breton dining-room pieces of darkest oak, dark as the original forests and the intentions of dwarves, taking up most of three walls.

At first glance this massive furniture seems as somber and overpowering as its German equivalents: the great carved chest containing not blankets, not skeletons or furies or vipers brooding on dragons' eggs, but the six or eight yard-long daily loaves of bread along with a hoard of tapioca, sugar, etc. and a smattering of broken coffee-grinders and other non-edibles as well, the French system of domestic storage being such as to give any American housewife an immediate nervous breakdown; the kind of cupboard you hide in in fairy tales, with its equally prolix table of contents; the giant sideboard surmounted by tiers of racks for the display of the best, that is never to be used, china. They are woman-killers all right, offering with their scroll and relief work an absolute maximum both of surfaces to dust and polish and of daily housekeeper mileage—hence the carpet slippers and the swollen ankles; it is as good as padding ten times around a football field to get out the few items needed for the simplest meal. Not that there isn't plenty of room for cabinets in the otherwise modern Hervé kitchen, only nobody has thought of putting them there. And it wouldn't be fair to the heirlooms to take away their function. Anyway to a longer view this furniture is quite, if subtly, un-Teutonic. Some lightness of spirit has been breathed into the heavy wood; it does not lean over you but stands upward and away, by a delicacy of proportions and carving and in the sparing use of brass. It is peasant style very tactfully refined, not to make you stand off and admire but just to give a feeling of comfortable pleasure in your skin and soul, which at the proper hour, if you were not the one doing the trudging, suddenly translates itself into a terrific appetite.

Still, a discriminating appetite, whetted by oysters whenever Monsieur chooses to bring some in; they are his own gold and he can afford to eat it, unlike the employee population. (However in the course of a year he gives away at least a ton of crayfish and such, whatever it has been his pleasure to catch.) These people are not burpers at their food like Germans, and of course not farters and vomiters like the old Romans; nothing so gross, as the furniture in its rather surprising grace and distinction might tell you. The same huge chest, cupboard, sideboard, racks, placed just where they are now, as in so many other well-off Breton households—for this was one of the autonomous arts that flourished throughout Brittany for a couple of hundred years—have presided over generations' worth of generally simple but serious and well-cooked, well-appreciated meals, hearty at noon, light at supper, and have had their role, as oysters and the lovely flux of the landscape do, in all that Hervé digestion. In three ways: by handsomeness, solidity, and tradition, having gone on radiating their message in all registers, from gourmet pleasure down to the stern reminder of the reverence due to food as life, through a good many disasters national and personal and sometimes a case or two of melancholia in the family, of the kind the present mistress is not entirely immune to, especially in winter when guests are few.

There is nothing of that in Monsieur, on the contrary. He is never idle, deeply loves the oyster business including even now the rigors of it in winter, even the uncertainties in a bad winter, more mildly enjoys his farm, has had a lifelong infatuation with that tidal pond of his; its every change of light and level, and it is never unchanging, speaks to him; it is like some great nebulous animal two acres big that loves him and loves to work for him and play with him yet keeps its own wild sea-secrets too. If it ever dried up or became polluted he might well die of grief.

But just now, in late May and early June, getting on to placing time, a faint trace of melancholy does show in his handsome weathered old features as he sits down, robust and momentarily grave with both the moment's unspoken ritual and critical expectation, to the noon meal. It really is not the same, buying his naissain. If only his sons, or one of them . . . From the bridge these days he can see Mme. T. in her gauleiter-type coat giving orders and worrying over her timing again, as well she should, poor prickly soul with her charmless restaurant charm that always makes him think of a cash register, though he has never had any complaint to make of her as a working neighbor; and the two chantiers the other way beyond his own now silent one, with the women making such a party of it. Everywhere he goes along the shore on his bicycle or in the car with his wife the white piles are rising, at Saint-Philibert and La Trinité and over at Le Bono where ten years ago there was nothing. Every Tom, Dick and Harry is getting into it now and indeed he wishes most of them well, those who are doing it in the right spirit, for he is more proud of his *coin* than of his own role in its unique production. He sniffs at every breeze as though forgetting that he is out of it. Oh yes, he still has oysters, but only the way he would up at the Rivière d'Étel or on the north coast, by buying them, not in the rightful way of Locmariaquer.

The sea is swelling with that precious progeny and although he is not an emotional man it is as though in his own body too there were about to be an explosion of new cells of life, he has lived so long with that cycle. Everywhere but that one place beyond the bridge, his place, the barges are drawn up ready to shove off, and all his aging body aches to be ready too, with nowhere to go, no tiles any more to take out; every handful of water bursts with the mystery; the days and minutes tick toward the flash moment that is to be caught, as by an army general in the field only not for human slaughter, moment of nature's big stunt and possibly our little earth's distinction for these recent billions of years—life, new life thickening the waters, and along with it the invisible accounting written on a wall of water, of the oysterman's luck for the year. Then the two or three weeks of waiting, until you begin to know . . .

There is a story of whiteness and the sea that is not on the glass-fronted Hervé shelves, nor in the Municipal Library of Vannes either—*The Narrative of Arthur Gordon Pym*, by Edgar Allan Poe. It might perhaps be better understood around Locmariaquer than in the sort of places where it is usually read; or perhaps not. The end goes something like this. The previously shipwrecked boy, or perhaps it is two boys, who during their adventure on a South Sea island have gradually and in spine-chilling fashion been learning of the native terror of the color white, under pursuit make their final escape in a canoe, taking a captive native along with them. The boat drifts southward at no great speed at first, then faster, at last vertiginously; for pace and other details, see original. The point and climax, and of what starts out as a rather ordinary boys' adventure story, the awful mystical height Poe has been getting ready to fracture your everyday soul and senses on, comes when the native, looking wildly ahead, lets out a ghastly shriek and falls unconscious in the boat—or canoe, whatever it is. And there before them as they speed more and more madly toward it, the young innocent civilized others see *it:* the wall of pure whiteness, into which we leave them vanishing.

AROUND the Basin of Arcachon some oystermen will tell you that they don't trust the Institut des Pêches, or Bureau of Fisheries, in the matter of when to put out their collectors. Although dealing by now almost entirely with Portuguese, with various differences in technique, the industry there still shares that much with the Morbihan. The tiles, which are the same shape though larger and stacked differently, have to be put out on time, and the Bureau is sometimes accused of not sampling the water widely enough to gauge it correctly. It can happen that an oysterman will find the larvae fixing thickly in his particular area or even that it is already too late, when the official word is still negative. Presumably the same could occur in the Morbihan, but by and large there the oystermen's complaints against the government are on other scores—antiquated legislation, some of it dating from Colbert, or the general insufficiency of government help to the industry, which certainly is on a pathetic scale by comparison with what went on in the Second Empire. For Doctor X. of the little Auray office there is general respect. There must be a margin of error as in anything, especially considering the huge extent of the bays, coves, estuaries, etc. that he has to cover and with only one assistant, but generally speaking his word rules on this crucial point and is found reliable.

It is a killing period for the two of them there in the lab. They are off at dawn every day after their specimens of water from here and there, and bent over their microscopes till all hours every night, just when everybody else in the oyster business is taking it easiest. Once the tiles are ready most of the workers are laid off for a while; both family-scale and big-boss operators have mostly finished what has to be done, and the big Oystermen's Cooperative building on another back street in Auray is more or less empty. In a little glass-enclosed office by the entrance the dignified President, himself an eminent oysterman, who will represent the Morbihan at the national meeting in August, gets off a letter or two and departs. There will be plenty of ticklish questions as usual at the meeting, to be washed down with good wines however and besides he enjoys the annual fray: price regulation, formulation of requests to the government assuming there is one by that time, but of course in Paris the same bureaucrats go on functioning anyway with or without a government, matters of expiring leases on oyster parks and granting of new ones, and so on. There is some politicking to be done beforehand but nothing much just yet, and the one salesman left out on the big floor has hardly anything to do. Leaving out boats and barges and such, the really big things, there is just about everything there you could ever need in oystering, from women's dark blue canvas pants and the beautiful fishermen's blouses up to the heaviest cable.

A lovely store, in which nothing has been advertised, nothing is packaged, no patronage is solicited, no brainwashing is done, no profit is to be made and therefore it is most unlikely that anything bought will break or otherwise go to pieces the first time it is used. Oh, the long-lost delight of this decency! For the general American public there is nothing left that begins to approach it but the small-town hardware store, where a nail is still a nail and had better be a good one, and there is apt to be a good deal of junk and vanity even there. Besides, in this place with its crude heavy counters and air of a warehouse, the aura of the single trade, the *beau métier* with all its sea-depths and adventure, hangs around every item, for not a bolt or rope or pair of gloves there is meant for any other purpose, and it is remarkable what beauty it casts over everything. Beauty depends after all on what you come from, what you are being cleansed and relieved of, and in the pass we are in nowadays an American lady of the buying type might be tempted to come away from this place with a batch of pulleys, the way her grandmother acquired a little replica of the Venus de Milo.

But no, that wouldn't do, would it? The beauty of all these honest things, aside from their fine conjunction of textures, is in their being together and being there, not somewhere else, in the above-mentioned association, in the simple appropriateness of it all. It is not to be bought; the poor lady will have to go back to the square, with its tasteless souvenirs.

Meanwhile it is party time in Locmariaquer. The schools are still in session but no real work is being done any more. There is a rash of amusements. The little one-ring one-family circus with its three scabby lions and long-winded joke-telling clowns and a horse with diarrhea and talented sad young acrobats, the son and daughter of the family, who hawk candies and take up an extra collection at intermission, puts up its tent for a one-night stand in the little square, really only for the bourgeoisie of the village such as that is, since the oyster-workers can't afford tickets. But the children get a good enough show from the tent going up and the lions roaring beforehand. And the Abbé's big picnics. And the big local fête of the year because every bourg must have one, also run by the Abbé, for which the two clerical schoolyards which turn out to be adjoining, the boys' and the girls', are opened into one; bagpipes, imported dancers in costume, Giannot and the other farmers in a parade of tractors bedecked with flowers, the main convent schoolroom rigged up with tables and benches for eating and drinking. It goes on most of the day and way into the night. Marie-Jeanne is one of the volunteer waitresses. The Abbé has had his family come over from Baud and his good rolling laughter rings now and then through the crowd indoors or out, contagiously but not in life-of-the-party fashion; nobody is captive to his jokes and good humor, people around him feel rather that he is responding to their own sallies, and so the fun multiplies. Marie-Jeanne, scurrying here and there, is not aware of glancing his way every time a burst of laughter eddies up around him, but does demurely know that her heart is warmed each time and that the job becomes suddenly tiresome when he is not in the room. She tells herself that everyone feels that way about his presence there, and up to a point she is right.

Another party a few days before. All the employees of the N. establishment, where Yvette works, have been given a banquet in honor of the wedding of young Monsieur N.'s sister. They were all at the church first, then the wedding party and intimates, the gentry, went to their separate banquet in Auray. The workers' party was at one of the two little hotels in Locmariaquer and very gay it was, with oysters and huge roasts and all the wine anybody wanted which was plenty before it was over. Yvette was ravishing, not so much from being in her best clothes, becoming though they were, as in her rollicking delight at being at a party without her husband, who works with her brother at the H. yard so of course wasn't invited. So instead of having to watch him get drunk she was the belle of the feast and star of the dirty phase, which she knew very well how to handle without being a snot or pretending not to be flattered by it—it was mostly good clean friendly smut; and ended up tight as a coot herself. She just barely made it home on her bike, and the children who had been waiting for their supper, before she had even turned in at the gate judged it advisable to go over to their grandmother's.

Then the word comes. The first tiles are to go out tomorrow.

AT the N. chantier and most others the workers go out all together with their load, on the boat or the barge itself. Yvette and the others are assembled at 4:30 A.M., ready to leave as they must in this case a half-hour before high water, to be at their destination just after the turn of the tide. The day promises to be fine and clear—not that that means anything but it puts everyone in a good mood. This is really the best party of the season; nobody looks sleepy, everybody is lit up by the sociability and the strange undercurrent of excitement to it, almost any sort of joke gets a big laugh. The voices pick up depth on depth of resonance from the water as they skim out in the dawn, wheels within wheels of sound, to sink and die far out in the bay. From the H. yard, on the other hand, Yvette's brother Yann the foreman sets out alone. Monsieur H. drives the rest to where they are going, the inlet by the bridge at La Trinité, in the car.

This suits Yann. He is the ungregarious and unmarried, the rather somber one of his family, given to few words and less laughter though easy enough to get on with. He lives alone at the chantier and goes to his parents' farmhouse at Saint-Pierre for the main meal every day, normally at noon but it will be in the evening for a while now that the tile-placing has started, because of the hours of the tide. He is very tall, slim, with crinkly blond hair and a muscularity somehow furious, as though some process like the gnashing of teeth went into building up all that strength. There is nothing either classical or typically Breton in his good looks; no one feature is narrow and his eyes are not narrowly set, yet his face totally lacks the moon-roundness of his beautiful sister's or the withered-apple version of the same, their mother's. It is long, intelligent, only superficially quiet, speaking of large inward forces continually at work. Since Marie-Yvette's suicide, love and marriage have been themes abhorrent to him, and what he picks up in the way of sex in Auray or Quiberon is nobody else's business.

The two murderous black dogs that guard the H. property, both house and oyster grounds including the parks, sniff around him as he gets ready to shove off. There have been no more thefts of oysters since they came and they would never hurt him or any of the other regulars around the place; still, they worry him, sooner or later there is going to be trouble from one of them. However, Madame wouldn't hear of getting rid of them . . . The great barge swings slowly into line behind him as he hits the channel and heads out, to where in lieu of horizon the sunrise throws up sherbet castles pink and lemon, slow fountains of architecture amalgamating sky and sea. The solitude elates him, and the night chill still in the air so that he is not ready yet to take off his heavy blouse. The parks, the ones he is responsible for and which are the only ones off that point, strewn with last year's naissain and farther out with the eighteen-months, are of course covered now. Nothing shows but the tops of the chestnut branches forming long tiers of fences out beyond the parks, in the manner of snow fences—a big expense and nuisance of that exposed location, where without them the roll of the sea would toss the oysters about too violently and tend to bury them. Chestnut is the best but even so gets eaten by the worms as the pickets do and has to be replaced every two or three years. For pickets Monsieur has been importing *azobé* from Africa for its worm-resistance, at frightful cost.

Out around the farther point now and goodbye to solitude, he has entered the procession coming from the Gulf, although on this lap the barges are too far apart for greetings or for any special maneuvering to be necessary. But he can recognize most of them and is slightly annoyed to find Mme. T.'s small boat, a third the size of his for the H. boat is also a dragger, next in line behind him; her son, who runs the restaurant now, is at the wheel. And just back a little way on the shore, among the scrub-oaks, Françoise will have been standing, watching, luckily too far to tell one boat from another . . . The Janec family have never told what they know, it would not be wise, nobody saw but Yann's and Yvette's older sister Léa, the ugly one, who happened to have a job in the restaurant for a while that year and at that moment, on the day before the suicide, happened to be setting tables with Mme. T. instead of doing something out back in the kitchen. So she had to see. Afterwards she said she had wanted to scream at Marie-Yvette for the love of God to go away, not do it, go in like that for the first time in her life and demand to see her father. But she couldn't make a sound and had to watch the whole scene, exactly as she knew it would be—Madame spitting at Marie-Yvette and calling her those awful names and kicking her until she was actually down on the floor crying, she who had always stood the straightest and proudest of them all, and then whipping her around the face with a table napkin of all things, because she didn't find anything else at hand to do it with. She called her a liar, blackmailer, whore, threatened to haul her in before Monsieur T.'s cousin the Chief of Police across the street.

A long time ago now, and Yann, who at the time could gladly have strangled the whole T. family with his own hands, knows that his mother was right. Not only that they must never tell, there was no knowing what revenge or bad luck it might bring on them, but that they must not judge either, although she had cried as bitterly as any of them when it was all over. They shouldn't judge, she said in her grief, in the torment of her compassion for Françoise, trying hard to mean it; meaning that that other one, fiend out of hell though she seemed to them and might be, was a wife and mother too and how could they know what she had suffered, to drive her to that; which of them could be so sure they wouldn't have done the same?—even though the only mother they had on their minds and hearts that night was Françoise.

Yann finds himself heading straight for the cliff, veers back in time and the T. boat passes, with no show of recognition on either side; and in fact the T.'s being from Auray and status-proud don't know his name or that his sister once worked for them, he is just Monsieur H.'s foreman. He has to look sharp now anyhow, turning in toward La Trinité, into the real thick of the traffic going both ways, all with the same white cargo. Here the jokes and greetings fly every which way; it is a regatta suddenly and even Yann is transformed. You could almost take him for a gay-dog type. He is well-liked for all his air of withdrawal and this is one of the times when the fact shows most pleasantly. An old woman in a shawl, a distant cousin of his mother's, throws a bouquet of dahlias at him; it goes in the water and she hollers to all in hearing that she'll catch him yet.

A couple of men friends chugging the other way wave a bottle of red wine and there is some bellowing of witticisms back and forth over that. The hundreds of pleasure boats of all sizes lining the little harbor and already in trim for the summer, although their owners will mostly not be there for another ten days, look unnaturally still and rather pathetic in all this swirl of festivity and activity in which they are totally ignored. Accustomed as they are to admiration and loving care, they seem to go suddenly brittle in this situation, like so many toys mysteriously strewn along the walls at a grown-up ball. The real mystery is how they will ever get out later, once the tiles are all placed, but they do; certain channels, hard to perceive at this juncture, are left.

Burst of exuberant hello's from the H. group on shore; Yann anchors and goes for them in the dinghy, and almost at once the work begins. The timing has been perfect. Ten minutes earlier the tide would still have been too high; now when the men jump overboard the water is up to their armpits, which is awkward enough but not impossible. There are four of them, including Yann: a young boy; Yvette's husband Jean-François, mercifully sober and not unattractive today; and the poor dim-witted middle-aged underdog always to be found in a crew, who gets the blame for everything and the worst jobs, known in this case as Poujade because of his rabid devotion to that political demagogue and fellow-Breton. That the real Poujade's bubble burst some time back doesn't affect this one of his followers at all; he goes on making the same speeches whenever he gets a chance, and by now his own real name has been generally forgotten.

By the time they are through, the water will be breast-high again, that is on the taller ones, which in the case of Poujade means to his neck, but there is no great sense of rush at this stage. The boat is anchored a little distance away and will not be boarded again until the end of the day. Tools, lunch, the stacks of stakes of two different lengths, everything they need is on the barge, where Léa and the two other women poke the stakes through the bouquets and hand them down to the men alongside. The placing of the first line of bouquets, and the first in each ensuing block, is crucial and ticklish, a job for nobody but Yann himself. Before that, there was the supporting fence to be put up at the outside limit of their territory, where the channel currents most endanger the tiles—two heavy stakes driven below the *vase* with a sledge hammer and a board nailed across. Against it the first row of bouquets has to be placed, not driven in but merely set with a slight push down, at exactly the right angles of tilt, toward the fence and toward each other, to withstand the stresses of tides and storms for the next six or eight months. If even one picket in that key row is out of kilter, the whole block of some thirty bouquets, or three hundred and sixty tiles, shortly to be coated with twenty, forty, perhaps seventy thousand baby oysters depending on the year, may be tumbled over in the *vase* and ruined. In fact it is not unknown for a single tile to bear as many as three hundred seed, but the whole mass would never average that, even without speaking of the mortality along the way.

It has to be done fast and be right the first time; it is not the sort of work you can correct later; the practiced eye and hand count for everything. *Un tracas de tuiles*, a "fuss" or "turmoil" of tiles, is the arresting term for the groupings of bouquets, by threes or by fives, spaced to allow enough circulation of water. On subsequent rows the others do the placing too, all but Poujade who gets bawled out if he tries to give himself a shred of dignity by sneaking a bouquet into place instead of handing it over to someone else. In friendly vein though, and he laughs over it himself. He is the one with no boots at all, just a pair of old shoes to squdge through the *vase* in. Nobody goes barefoot in the black slimy stuff, full of cutting shell-fragments and worms, but the only one of the crew with proper waders is Jean-François, spoiled child that he is—they must have cost him a month's wages, or perhaps Yvette paid for them. He looks and clearly feels quite superior keeping dry like that when the others are of course drenched. Yann and the boy trudge back and forth in rubber boots full of water.

But it gets easier. The water sinks to their hips, their knees, only the sun becomes bothersome. Léa and the two other women are sweating as they work through the walls of tiles on the barge, keeping them moving overboard. They need skill for that too, as well as physical endurance; a woman not used to it would be exhausted in half an hour and would not be much help anyway, because of the difficulty of getting the stakes through the bouquets. The stakes are not machine-cut, after all, but chopped from branches and it is a rare one that is straight and thin enough to slip easily through the opening between the twelve tiles.

Mme. H., for one, could never do it and wouldn't dream of trying. For that matter her husband has never made or placed a bouquet either, but this is the anomaly of their big-shot position. They are one in a thousand. The other groups working along the sides of the estuary that morning are all bossed by the proprietors themselves, and the crew may be just the family. Some are on home ground, with their chantiers right there, but even so they must have a barge; it would be impossible to carry all the tiles out across the *vase*, except for a very small operation, staying close in, or for the kind of pickup business possible in such a prolific spot. For that, strings of broken tile fragments, something like the *chapelets* or "rosaries" of slate or shells used at Marennes, or almost anything, may be put out as collectors.

The freak object on the H. barge that day is a collector of swirling plastic, one of Monsieur's experiments, to be put out along with the tiles. They all have a laugh over it, it looks so ridiculous, but mostly the laughter has passed now and the whiffs of gossip that went so pleasantly with the motions of the work. It is late morning and burning hot, the tide is nearly down and the white mass looks hardly dented, although in fact several blocks have been placed and the barge moved over twice. The whiteness in that light is dizzying and seems to have a causal relation to the eerie spell of stillness over the whole harbor. Not a boat moves in the channel; the pleasure craft lie flopped on their sides like dead fish. The hum of vehicles now and then on the high bridge sounds very far away, as do the voices of the tile-placers in the other concessions, as though the high poles with branches on top marking off the territories one from another actually marked a sound barrier.

Indeed a psychological barrier has gone up. The pressure is on, the pace of the work stepped up, the group is intent only on its own doings. It is the time for tempers to run short, and for poor Poujade to get it in the neck. Naturally he will be the one; it seems almost that he does the stupid thing on purpose, knowing that that is how he must justify his existence, by giving the others a vent for their rage. He drops a bouquet, soiling it badly, and as if that were not heinous enough, in retrieving it leans heavily against the stern of the barge, which swings over and becomes stranded right across the line of work. Shouts; curses; he is called every name in the book, not only by Yann, who as foreman is entitled and even obligated now and then to spew forth the mighty wrath that lies hidden in his judicious and frustrated nature, but by Jean-François equally, who but for this blessed interlude, and for all his virtue in keeping sober, might have to go through the whole day without anybody noticing his heroic stature or even without feeling heroic himself—and when he can't be a hero he is apt to feel like a worm and consequently to get drunk or beat up Yvette or both, because on such days everything seems to remind him that he has been fired by Monsieur H. half a dozen times for drinking and the next time might be the last, also that he is married to a strong and beautiful woman who is more often than not disgusted with him no matter what he can still do to her in bed. Well, he can do plenty, in and out. How she'll laugh, how her flashing eyes will caress his great muscles when he tells her this one! After all he is as tall as his brother-in-law and of far stronger build, he could practically throw that fool Poujade out into the channel if he had a mind to; not that Poujade is any weakling although at the moment, with his downcast eyes and bumbling explanations, he seems to be imploring them all to throw him anywhere they like.

Even the boy and the women are railing at him. "Dame, quel imbécile!" "Nom de Dieu, sticking us in this ass-hole with all these tiles . . ." "You want the whole *coin d'Auray* to be laughing at us, espèce de . . ." "Push, for the love of God, idiot! Not there! Here!"

And in the nick of time, in the last inch of water, they do heave the still half-laden barge back where it belongs, glued and clinging to the *vase* as it is and with their boots gluing deep with the violent effort of their bodies. Jean-François really is, in this contingency, rather a hero, out of the sheer fury of his determination to have the strength of ten and be able to tell about it afterwards. But all four have had to be somewhat superhuman, and lean panting on the side, even beyond imprecation for a minute, when it is over. The women could have done no good in the sandals they are wearing, so stayed on board.

Then back to work, even faster now, not to be overtaken before the barge is empty. The curses and other reminders to Poujade continue in little reminiscent dribbles for a while, but only serve to point up a mood of refreshment and good humor of a deeper order than in the early morning. Often enough before and in many fashions, august or trifling as it might be, the sea has given them this largesse of a chance to come through together, not to be outdone by it, and of a bond beyond any ordinary one of people working together. The sea, and Poujade—and the former in this case was just black mud. Never mind, the little drama has picked them all up very nicely, and the real hero was not Jean-François or anybody else six feet tall, but Poujade, and he knows it; without him they would all have been grumbling by two o'clock. He is positively coy for the rest of the workday, chipper as a chipmunk and bubbling with little jokes which oddly enough are appreciated for once.

It is done at last, the car is waiting, the tide is up. Yann stands on shore a few minutes conferring with Monsieur, who in spite of being cut off from the manual side of the work by taste, wealth and the scope of his business, nevertheless has the vast accumulated knowledge and uncanny instincts of the born oysterman. It is true that he made out badly in the deep-water experiment—the spot was ill-chosen and the preparations inadequate, but he hasn't given up the idea, and on the traditional scene nobody is shrewder; when he orders some measure against the judgment of his foreman and everybody else, it usually turns out that he was right. Tall, lithe, his youthful skin still looking almost cherubic above a walrus mustache that ought to look suspect but doesn't, he embodies the kind of deep natural and totally male authority that even such types as Jean-François find it hard to resent. Poujade with all his second-hand rabble-rousing would be embarrassed to tears to have to push him over, though he wouldn't mind hearing that Madame had been in some little "incident of the road."

The tops of the bouquets, theirs and all the others placed around the river-mouth harbor that day, are still showing. It is just a speckling now; the main whiteness of the Morbihan is still on shore. All is well. They agree on the area for the next day and Yann heads the boat homeward, taking the boy with him this time because the barge may be more capricious now that it is light. No procession at this hour, only three or four of the dark silhouettes to be seen at a time, spaced out across the sheen of the bay, each oysterman returning when he is ready, within the possibilities of the tide. If it had been souls they had taken out that morning, in their white packages, you would say there was neither grief nor relief in this return journey, but merely a fitting in with the laws of life and of a job to be done by generations of boatmen till the end of time.

The two of them eat and drink a little, quietly, exchanging few words. It is no time for blowing off, or being tired either; you have to just dull down when you can, and be ready for the next push. The others are already waiting at the dock, and quietly, as soon as the boat is moored, they begin loading the barge for tomorrow. In all, they will put out 250,000 tiles.

THE most famous incident in French oyster history occurred in 1868 and had better be mentioned, because somebody would be sure to take umbrage if it were left out. It has to do with the introduction of the Portuguese but in terms of economics and biology has a large bearing on O. edulis too.

The plight of the latter in the Arcachon Basin in the 1860's, at the height of Coste's work in saving the species and just as he was finally spared from all the crises elsewhere to set up his model "farms" in that region, brought on another, in a sense rival development. Hearing of the quantities of cheap oysters available at the mouth of the Tagus in Portugal, the oystermen of Arcachon decided to import them. Authorization was granted and they began bringing in between 25 and 30 million a year. They were for cultivation only; no natural beds were established in the bay. It was a purely chance occurrence, of the kind that can win wars and cause the great breakthroughs in science, that changed the whole picture.

A ship named *Le Morlaisien*, loaded with Portuguese oysters, was driven away from the straits of Arcachon by a terrible storm and took refuge in the Gironde. The cargo being by then putrid from being out of water too long, the captain, one J. Patoiseau, who was to earn through his garbage problem a peculiar immortality, dumped it in the river-mouth where it soon turned into a gold mine. Apparently only a few of the oysters were rotten, the conditions were ideal, and as there had presumably been some hundreds of thousands of oysters in the shipment, an enormous natural bank was established—the first on the French coast, and still prosperous now. The exact place of the incident was all-important; nobody in France at the time knew enough about the habits of C. angulata to have chosen it on purpose. Another key factor was numbers. A few hundreds or thousands of oysters are not enough to set themselves up on a permanent basis. Too many of the larvae are dispersed by currents; only a vast concentration has a chance of survival. So it had to be a pretty big ship, and the smell had to become too offensive to the captain over a spot where the right temperature, salinity and bottom would prevail not just for a run of lucky years now and then, but chronically.

There are a few other such spots along the coast, one at the mouth of the Charente, to which the strong currents out from the Gironde eventually carried enough of the Portuguese progeny to form other banks. So the new species was in to stay, as it might never have been, or not until much later, if that particular ship had been able to make port and had dumped the cargo there, because it seems that the Arcachon Basin is not a good enough place for it. Occasionally when there has been an abnormal amount of rain in summer the Portuguese has reproduced very well there, but as a general rule there is not enough fresh water at that time of year to permit the eggs to develop. You would never suspect this from a casual visit to the area. To the tourist eye it looks like an oyster heaven, with its parks all the way around for some thirty miles to Cap-Ferret and its thousands of tile collectors. The tiles are not on stakes but in heavy wooden cages, since the seed is left on until the second year, unlike the *plate;* that is because nobody bothers about the shape of the Portuguese and it is too inexpensive an oyster to warrant the labor of moving the tiny seed. If there are collectors there must be larvae; and there are, but not enough. The millions of oysters in the parks, concentrated though they are there, don't grow old enough to be proper parents, and the parks are in zones that are too salty anyway, though excellent for cultivation.

At least, they were excellent. In the last two or three years an obscure ailment has striken them. The death rate is very high.

Out in the Arcachon parks at low tide you see gaping shells everywhere, and the survivors are abnormally small; hardly any are growing to what used to be considered marketable size. There is talk of new factory wastes, especially from a certain wood-pulp installation that should not have been permitted on the Basin, and other contamination, but the real nature of the trouble has not been officially labeled. In any case the business is almost all in Portuguese, with only a small proportion of *plates* as a sideline.

The switch took a while, in spite of the *Morlaisien*, reaching its present aspect only with the disastrous disease of O. edulis in 1921–22. For a long time the Portuguese was considered the oyster of the poor, the "democratic" oyster; few people were visionary enough to believe that anything so gross in looks and taste would ever sweep the market. But the attractions were great. The seed, in the 1870's, cost the oystermen literally twenty times less: one franc a thousand, versus 20 francs a thousand for the *plate*, so people with little or no capital began to flock to the business. The work and the hazards were, and are, in about the same ratio. Less sensitive to sun and cold and so able to survive more frequent uncovering, it can be cultivated closer to shore, where along with other advantages, the suffocating weeds are less and crab-prevention easier. In general there are far fewer motions to be made and far less space needed.

Most important of all there is very little chance with this species, in contrast to the *plate*, of an oysterman's being altogether ruined by an abnormal turn of climate, whether of temperature or rainfall. The two most terrible winters of the last century, 1869–70 and 1962–63, have given dramatic evidence of the fact, but it applies to much commoner aberrations.

With time it was discovered that even the lowly Portuguese was subject to a good deal of refinement if treated properly, and a certain number began to come in for much the same sort of care as the elegant native. Even so in the last stages of the absolutely top-grade product, as in the *claires* around Marennes, there there will be ten or twelve Portuguese to the square meter and only three or four *plates*. Although the reasons are still not entirely understood, it is a well-known fact that no oyster can be of the best, or be at its own potential best, if it has lived all its life in one place, no matter where. That is one of its several points in common with humans. So there came to be quite a price differential in grades of Portuguese too, depending mainly on where and how often they had been moved. The price of the best of them is about two-thirds that of the cheapest "Belon."

These are the oysters, the only ones, the summer tourist finds say in Biarritz or Saint-Jean-de-Luz or anywhere inland, away from the production centers. Once you leave the Arcachon Basin heading south along the Atlantic coast there are no more oyster habitats until you get to northern Spain, site of the last natural banks of O. edulis now surviving without human assistance. Everything is inimical to them in between, the sand and salt of the Landes as much as the rock cliffs and fearsome currents of the Côte Basque. It is not "the season" for Belons, they will tell you along there, meaning that they are not shipped at that time of year. But even at Arcachon where there are still a few, the less flossy restaurants will serve only the one kind, the Portuguese, and the waitress won't even know them by that name; they are just "our oysters," *les huîtres d'ici*, as if there had never been any others there. If she has heard of Montaigne she probably thinks that is what he ate too. The local Society of Many Learnings ought to get hold of her. Of course there is no oysterman that ignorant; for that matter it is a rare oysterman in France, however unschooled, who doesn't know the name of Coste. There is always a sprinkling of *plates* on their tiles and they can tell the difference, as a plain mortal hardly could two years later, when the shells are the size of a newborn baby's little fingernail. But mostly what they say about the *plate*, both there and at Marennes, is strangely familiar—"Too much trouble"; "Why bother?"

There is also a sprinkling the other way around, that is of Portuguese on the tiles of the Morbihan. If a person wanted to go crazy, there would surely be no nicer way of doing it than by taking up the study of coastal currents. Of course there are many other designs in nature that would do. But for the methodically tottering mind, not the compulsive, rushing type, coastal currents would be sublime. They have everything, including a certain amount of the unpredictable; not too much though. For instance, it is not enough for an oyster that a river should issue to the sea, even with other appropriate conditions. To allow a good bank to form, the fresh-water current, which will be of fairly predictable strength and directions, must collide with coastwise sea currents so as to make an area of swirl, something like a large benign whirlpool. Without such a centrifugal force, to concentrate the larvae in sufficient numbers, they would be carried away in all directions and do nothing for the survival of the species. The addict can go on from this oversimplification to the further factors, of tides, seasonal changes, etc. that will bring him a little nearer to the whole truth, e.g.: "Owing to the rotation of the earth the flood tide swings to the right, hence the natural oyster beds at the mouths of most rivers are usually much more extensive along the left-hand bank of the river. In a large body of water such as Long Island Sound, Prytherch has shown that the flood tide turns toward the Connecticut shore and that the natural oyster beds occur where this flood tide enters the brackish water at the mouths of streams." (T. C. Nelson in "The Boylston Street Fishweir")

In the case of the powerful Gironde, and to a lesser extent the Charente and other streams of the central French Atlantic coast now colonized at their mouths by C. angulata, there are among other phenomena certain lines of motion northwards, and of the billions of oysterlings swept that way a few are bound to last out the journey and fix wherever they can. There is also some progeny from the parks in the rivers of Étel and Pénerf, the two places in the Morbihan where culture of the Portuguese is allowed. Alongside the big postwar bridge over the Étel, region of awful bombing in the past that is now receding so fast, you can see the two species side by side and cultivated by the same family, as in innumerable other places both north and south of the Morbihan—the big squares containing the Portuguese, spread so thickly in this case the hard gravelly mud is nearly covered with them, close inshore, and those for the *plates* farther out to the edge of the channel, as they must be. But that again is just cultivation. Neither species is raised there.

A question in biology at the moment is whether the Portuguese could be raised that far north. That stray individuals make out is insignificant; efforts to establish beds in England have failed, and to date there is no real bank farther north than Pénerf in the Morbihan. Another question, source of furious agitation in oyster circles, is whether it would kill off the *plate* by using up the food supply, whether naturalized in the region if it can be or by an extension of the zones of cultivation. Evidently it would not, at least it hasn't yet in the Marennes region where the two co-exist, but the native naissain business is so infinitesimal there, compared to this little slice of Brittany where it is everything, that that is perhaps not very indicative. The question in economics, meaning mass psychology, is the same, whether the cheap commodity drives out its betters, but that probably depends on which side of 1984, if there is such a year, you are considering.

DAY by day the white masses on shore diminish, and reappear in their new configuration; reappear, that is, half the day, having taken their place in the rhythm of wash and exposure without which one is not to imagine any face or flower here, and disappear altogether in the neap tides when low and high are scarcely distinguishable. But by that time in the cycle the job is finished. Earlier, from the bridge of La Trinité for example, you can see the hopeless-looking, antlike procedure day after day, and within a week the incredible results. The whole inlet is becoming fringed, to a distance of a hundred or two hundred yards depending on vagaries of the shore, with the white stacks whose intricate arrangement could never be guessed from that distance, and the first-time travelers and summer people who are beginning to flock in, if they happen to notice, as they don't always because of the human capacity to miss everything when you are looking for something else and most people nowadays are mainly looking for bits and pieces of themselves around the world, will ask, "What on earth is all that?" Somebody in the party says, "J'sais pas," or "Something to do with oysters," and they move on to the great field of menhirs, where the President of the Society of Many Learnings, having finished liquidating his miserable store, is swatting at giants as he gives his first lecture of the season, or to the jampacked beach at Carnac. The President, at the end of the tour, will receive tips like a bellboy, and also hawk what copies he can of his learned treatise on the local megaliths.

For Yann or Léa, if something should take them across the bridge on their motorbikes, there is a deep secret proprietary tie to their own section down there, beyond speaking of, beyond thrill; their commitment is too habitual for that. To a person who had set out the tiles only one day, or one season, it would be thrilling. "Those are ours, those, right there!" he would want to cry out, hardly believing that from such aches and drudgery should have sprung that perfect work of art. Poor foolish neophyte, but not far wrong. The tiles *are* what he thinks, static only to the static mind and unbeautiful only to the unseeing; their beauty is all motion though they stand motionless, and is in process now; they are launched, they are out on their own farther than whaling ships, to engage in their white secrecy with all that we least know, and in that long awesome encounter the work of art is becoming. So why not speak of the joy of the creative act in this operation? That is all any of us is doing—trying to put out our tiles right, to catch a few oysters.

Yvette and her brother and sister and husband would laugh, that's why, if they weren't too polite. Just the same, in some version that any artist or true fisherman would understand, it is part of what keeps them there; it has a part in the curious play of suggestions, among shadings of tenderness, pride, and being in something for life willy-nilly, that their voices give to the word "métier." Anyhow, whatever these truths, they all go underground now. The real life of Locmariaquer, as of other coast towns around and most of the Gulf with its islands and nearly all the rest of the French coast from Belgium to Spain and Spain to Italy, is squashed out of sight by the summer invasion like a sapling before an army tank, and as though in sympathy with the reserve of the inhabitants, the tiles gradually become coated with *vase*. By August they are hardly noticeable at all. A few bargeloads, brown and sprinkled with a catch that is still scarcely visible, are to be seen as early as July going back where they came from. A few oystermen with appropriate grounds place the bouquets over again at that time, to have them nearer at hand or to put them in stronger currents; in that case they are merely set one on top of another with only a few stabilizing stakes the first time. But that would be economical only for a rather small number of tiles and is not the general procedure.

On shore, at the chantiers, there is one startling phenomenon. The large rectangular masses that have appeared there, so large as to dominate certain stretches of the shoreline almost as the white ones did earlier, are a shocking, stygian black. They are the newly tarred *caisses*, or carrying boxes, with wire bottoms and the wooden frames extended to form a pair of handles at each end, being put in shape for the work in the fall. In stacks of a dozen or so, they are quite funereal, in an ancient sort of way, especially as the great stone basins at the high-tide line are mostly empty now and look weedy and unkempt, and there is such silence around the yards. There are a few trayfuls of oysters in some of the basins, for the proprietors' own use or for sale to local restaurants, and now and then if it has been hot a boy will be pouring buckets of water over them. What has come to life, and as by a mechanical signal precisely on July 1st, is the pleasure boats of all sizes and various pretty colors of sail, for they are almost all sail. It is their turn now. They have sprung out of their forlorn and useless state, and for exactly two months will reign in regal flippancy over all the waters around.

LET'S face it, there are too many people. The shores of the temperate zones are simply not big enough to hold them in July and August.

In the region we are talking about, Carnac-Plage and Quiberon are too horrible in those two months to dwell on; even so, of course, they are nothing like Biarritz, La Baule and so on, the really good places. Locmariaquer, having no good beach accessible by car, has no good hotels either and so is rather a backwash; it merely increases its population a thousand per cent. The old German pillboxes sprout awnings and stovepipes. Cows are moved over to make room for a huge "camping" with prefab toilets, coin-in-the-slot iceboxes, other people's radios. Fields and streets swarm with blue- and pink-uniformed children being marshaled about according to that great French institution, and it really would be that if anybody could breathe, the *colonie de vacance*. Some are in big dormitory buildings that exist only for that and are closed the rest of the year, but the nuns' schoolrooms, crammed with cots, are housing a *colonie* too. The regular nuns, all but Sister Sophie the nurse, have gone for the summer, and another set of nuns is in charge. The Abbé is about to squire his own *colonie* of some thirty local boys all the way across France in a third-class carriage to Alsace; along with their sheets, shoes, etc.—and if they can't afford a good pair of shoes they can't go, because their chief diversion is going to be walking fifteen or twenty miles a day—each boy takes along his own butter for the month; their families don't trust foreign butter. The foreign element in town and in the section in general is all French, with few exceptions; the department isn't either toney or picturesque enough to draw many true foreigners, like Finistère.

The Janec and Giannot farmyards, meeting at some vague line in front of the little chapel of Saint-Pierre have become, in the most dreamlike sequence of all, a parking lot. Certain vast buildings one never noticed among the trees and in the huddle of ruins have suddenly made their appearance and are now occupied, to put it mildly, one by a Monsignor from Paris with his retinue; another, a few yards along toward the beach where the bonfire was, turns out to be the family *maison de colonie* of a big trade union. And over by the dyke the pink mansion is open too; it was rented for a huge sum to a disgruntled family from Nantes who have no use for the old oyster basins and have discovered that the plumbing is defective, in fact doesn't work at all. There doesn't happen to be a good hollow tree around; if there were it would have put forth curtains and an inflatable plastic swan. A little made-over farm building at the bottom of the Hervé orchard is housing in its three rooms three families, total twenty-two people.

And so on and so on. Busses all day in the *place*, excursion boats around the Gulf, folklore festivals, another circus or two with plenty of customers this time. The little stores stay open at lunchtime and far into the evening and still are bulging every minute. You wait half an hour to buy anything, and the baker and the butcher and their wives and Marie-Jeanne and her mother have dreadful circles under their eyes. They simply can't keep up with it all but they have to keep trying, at four in the morning still doing their accounts and opening again at seven. The baker is the worst off, but over Bastille Day and August 15th, the other big time, Monsieur Madec the butcher too doesn't go to bed at all for seventy-two hours; it takes him all night to get the meat ready for the next day.

Yvette is working for a family from Paris, no connection with the other family from Paris in the old white remodeled farmhouse nearest Françoise. These last, in the house where she and her mother were born, are in fact Françoise's landlords and allow her to go on living in the hut in return for watering their flowerbeds and helping out in the kitchen when they are desperate. The lady of the house is quite often desperate, as with children, grandchildren and other guests there are always twenty to twenty-five people in the house, all of exceptional charm and good looks, and all from the infants up giving orders to one underpaid domestic. She is a servant after all, and a Breton besides. In the case of Françoise it goes farther, to the accompaniment of many kindly words and little gestures of charity. It rather pleases them to have such a pitiful case under their wing, not that they know much of the facts; somebody did tell them once that she had had a child long ago, who died, but it tends to slip their minds. They are sorry for her because she is poor and crippled and wall-eyed, and because they consider her an imbecile. "She is rather *off*, you know, poor thing"—tapping their heads. It is true that she is not very good at washing the dishes, of which there are seventy or eighty to a meal without counting pots, to be done in one tiny sink and stacked on the floor meanwhile for lack of a counter, and she does make poor time with her stick in the ten-mile run, as it adds up, between the kitchen and the beautiful Breton chest, sideboard, etc., the family having of course furnished the old place in the provincial style. So it is "Dépêche-toi, Françoise," and "Nom de Dieu, qu'est-ce que tu fais là," most of the time on those occasions when she is called in to help. But they give her a little cash too now and then, and pay her for her eggs. It is more money than she sees all the rest of the year.

She doesn't mind, it is all too alien and incomprehensible, it passes over her like a freak storm. The regular maid for the summer, a fellow-worker of Yvette's at the N. yard, gets furious and swears she will never work for any "Parigots" again, although Yvette doesn't make out too badly with her Parisians, who are not the titled kind like those nor particularly well-off either. Their arrogance is of a sort she can understand and ignore. Françoise ignores without needing to understand. She is one person in Locmariaquer to whom the summer brings, aside from the excursions into the whirlwind down the lane, a mellowness of life, and a relief from some of the frights of the winter. To go out fishing with her knife and basket becomes almost a pleasure, and she stops thinking every night that she may find her mother dead in the morning. The old lady even seems to hear and see better in the warm weather, and on a good day, with diapers and the huge headshawl tied as on an effigy to hide the crumpled semblance of a face, can be deposited for hours out under the trees and hear all the strange voices, for at high tide their little yard, which is really a living room, becomes a thoroughfare to the beach. A modest beach, of stones, good for children, the little bay is so calm and shallow so far out; the kind of people who seek it out are friendly as a rule.

Besides they are not all foreigners. A lot of them are just from the bourg, or Auray, and will stop to visit, and there are others who come on purpose to visit. What frightens Françoise in the winter is her mother getting sick and there being nobody to call for help without leaving her alone, and then not being able to read the prescriptions. Now there is company several times a day, and she has a large supply of caramel candies left by both the Abbé and the priest in Auray, to give out to children. Her mother likes them too and lets out a little gurgle of satisfaction when one goes into the soft palpitating chute between her lips; it has to be stuck far in or it is liable to plop out again. But the mother, more than Françoise, is the one who likes to discuss politics. From every appearance of having drifted off into the slumber of a water-soaked log, she will emerge with a startling croak to make some monosyllabic pronouncement or to ask a question about the farmers' strike, and she accurately remembers the names, picked up heaven knows how, of quite a number of kings of France and even some of their successors, such as Robespierre and Pétain. A bunch of no-goods in her opinion, but she has always maintained, as her father did, that Anne of Brittany was doing what seemed best at the time in marrying two of them. Who can tell, there might have been even more wars otherwise; still, she would rather see Brittany separate, and often asks her visitors how the movement is going. "Not yet? Well, next year perhaps." She gropes in the air, her hand a slow tentacle seeking contact, for another candy, and then as though her mind had been recalled to the business of her own decomposition, the living part sinks again into the inert.

Françoise, more the poet type, speaks of more personal and observable things, her chickens and acquaintances and matters of the sea that she is in a position to watch, but prefers to listen, and in the course of a summer a vast storybook of happenings is brought to her there on her dais of dirt. To those who know her well, and it is surprising in summer how many they turn out to be, it somehow is that. Crooked-limbed, outside her crooked door, among the fleas and shells and chicken-droppings, with her grieving right eye and great swimming left one and only the little white cap in place of a coiffe, and only the sweetness of her smile in place of any claim or presumption, she does somehow hold court for those two months. Her sorrows alone would never account for it, although they have taken her beyond most people's range of vision and given her a certain aura of the sacred. The power is rather in her shy and vivid intelligence, which never seems to come out with anything very remarkable, but just goes flickering and stabbing through the mists here and there and leaves people, omitting the Parisians, with a feeling of having been rather dense before.

In late July Yvette's children bring her a potful of fine juicy blackberries from the hedges over there. You wouldn't think there could be bushes in the world so laden they wouldn't be stripped in a day in such circumstances. These are definitely enchanted; more tourists, more blackberries.

Then in August there is another item in the secret life of the village, which touches Françoise, for she has always been fond of horses. As a little girl she used to love to take care of her uncle's work-horses, and would be quite sarcastic about the cows. In late years, passing on her daily trek for the milk, she has seen a number of horses come and go in the field between Yvette's little house and Saint-Pierre, but never any that gave her as much pleasure as the four this year. The bay and the sorrel especially with their flying manes and tails, always leading the others in their wild capricious gallops, in circles like dry leaves or streaming to a smash stop, all shudder and incredulity at what they see, or at being stopped by any material thing, at the edge of the salt pond, seem made of something more than mortal. Whatever it is instead she may half feel that it is not unfamiliar to her, in a common-garden sort of way. There are many such spirits in the land, as there are voices in the sea, though not often in the shape of horses. Of course she knew how it would end too. It happens one evening as she is going by with her pail.

The truck is there across the field by the little white house, and Yvette and her family and the man are at the gate. There is a lot of smiling and joking and perhaps discussion of further business, for the man owns the house and this is how Yvette earns her rent. The horses have already had halters and ropes put on them, and the little girl Alice has her arms around the neck of the bay and is rubbing her cheek against it. She never got that close to him before but he is unnaturally subdued, like one of those articulated toy horses in whom somebody has just broken the elastic; now he can just make a floppy pawing motion with one hoof, and his head that was all fire and sunrise sways back against the child in dismal gratitude, nuzzling her. He does flash out once more at the start of the ramp, rearing straight up and striking the air with his front hoofs, in magnificent objection to man and all his universe, but is beaten on in by a couple of blows of a club on the back, and the others follow. The truck turns, exhibiting on its side a gilt picture of a horse-head and in big letters, BOUCHERIE CHEVALINE. There are three such butchers in Auray; the prices are only a little less than for beef.

The little girl and her brother, in the novelty of having the gate open, prance laughing into the field, playing at being horses. "Pauvres bêtes," Françoise says to the Janecs, standing a while in their doorway as usual after filling her pail. The rest is in Breton. "They weren't lucky. They lost their light." "Like us," says Mme. Janec. "Yes, that's true."

BUT most secret of all is what is happening on the tiles. Secret from mortal eye, and more than that, for we are prowling around the last question, *the* question unless the origin of life and the purpose of life are taken as two, and that remains shut for all our starfish pull upon it. Perhaps we are not yet quite wicked enough to be allowed so much; certainly the good fairies, who are stars in the sky, are outside of such inquiry. The answer is in their being. But on a modest factual level certain phenomena have been exposed in laboratories, so it is possible to admire, a little more fully than in ages past, the marvelous mechanism which is the oyster. "Greatly more complicated than a watch," Thomas Huxley called it, with true British understatement.

In the slimy deeps of time, when nothing was here that we would recognize but sky and water, there was possibly or probably, so we are told, a common mollusc ancestor. At least this is a current hypothesis in reputable quarters. It would have been a humpy sort of thing, with a head perhaps furnished with tentacles, a "foot" something like that of the snail, and around the body a layer of tissue or mantle, with the function of forming a rounded and originally horny shell; some eons later the shell becomes calcified, through the infusion of certain salts in the environment. There are some well-developed muscles, and we may perhaps imagine the head withdrawing into the shell when it wants to, like a turtle's. In other words this is already pretty far along in the scheme of things, but on such a casual promenade it would not be wise to look further. Rushing back toward home, i.e., the present, and leaving out the five other mollusc classes represented by the squid, the snail and so on, we see the shell eventually flattening and as though bending, until it consists of two plates connected by a horny ligament, to form what we know as the bivalves. At some point the mantle becomes attached to the shell near its edge by muscles, which after some more trials and errors fuse into two adductor muscles, front and back, with the job of closing the shell or relaxing to open it. A complicated system of little interlocking teeth appears near the hinge, still kept by most bivalves but not the oyster, to keep the valves from sliding sideways when closed. The foot, to be seen now in the scallop although no longer serving for locomotion, persists for a while; the head soon vanishes, as ours would too if it were boxed in like that; feeding is presumably by lips or labial palps, which some bivalves will later push out into the proboscis, like a retractable elephant's trunk, but most will reduce to little flaps for sorting; and in the case of our friend, along with a few of its relatives, the habit of attachment or "fixing" causes the hind part of the body to grow at the expense of the front. In the mussel this leads to, or goes along with, a growing disparity between front and back adductor muscles, a condition known as heteromyarian. The oyster, among others, in the course of evolution drops the front one altogether.

This little résumé, respectfully culled from *Oysters* by C. M. Yonge, is given to indicate something of the stupendous work performed by the oyster larva in the ten to fourteen days of its existence as such, that is between leaving its mother and fixing wherever it does, thereby ceasing to be a larva; for what it does is in part a job, nearly as hasty as this but written in living matter, of historical recapitulation. In some quite astonishing fashions the microscopic speck, if it lives to finish its story, re-creates the history of its kind. But first we have to show the mother at work, and anybody who calls her sluggish merely because she can't displace herself had better think again. Furthermore the spawning process is unique; no other bivalve perpetuates itself in quite this way.

In all oysters, the sperm when fully developed go into the exhalant or out-breathing chamber and from there to the sea in the usual water current. The female phase is far more taxing, the eggs being driven by elaborate muscular machinery back through the whole corridor of gills, against the normal current, into the chamber designed for *in*-coming food and water. It is a hall of a thousand doors, opening the wrong way, like a scene in a Cocteau movie; the proper exit has been closed. For all this to happen the temperature, for Ostrea edulis, must be at least 59° F. and for the health of the larvae preferably a little higher. Taking a colony as a whole, since some individuals are busy changing their sex at any given moment, it goes on over a period of several weeks or in warmer waters several months, but with a definite early-summer maximum from the Morbihan north. The rate of sex change also depends on temperature. It is a great deal harder to turn from male to female than vice versa—the production of sperm takes less out of you, so although off the south shores of Brittany an oyster starting the season as a female will most likely be a parent twice that year, the one starting as a male may have to wait till the next summer to function again unless it is unusually warm. In the Mediterranean several switches a season are possible; in Scandinavia only one performance a year in either role is common. A relation between spawning and the phases of the moon, that is the spring tides of both the full and new moons, which was remarked on by a few oystermen in the last century and denied by others, seems to have been borne out by the research of the Dutch biologist P. Korringa. Although there is apparently no such perfect subjection to the moon as in many other marine animals, he did find the greatest abundance of larvae, at the site of the study, "about ten days after full or new moon" (cf. *Oysters*). But it is time this sort of information was kept out of the hands of family planners; it really should not be published.

The Crassostrea mother, such as the American and Portuguese, having driven her progeny into the cul-de-sac, gets rid of them at once, through a series of contractions over the period of an hour or so. The motive in doing so against the stream so to speak is not merely to be ornery and different, this being the oyster's unique accomplishment among bivalves and a rather astonishing one to contemplate. However it evolved, the point is that that is the side where the shell opens widest, also greater muscular force is mustered for the last push, and the helpless eggs need to be thrown out to their fate more strenuously than the sperm, which can swim. The Ostrea mother takes in sperm along with her food and drink, and eight days later pushes out a set of babies excellently equipped with a sail called a velum to swim and feed with (hence the term *veliger* for the creature at this stage), a fringe of lashes or cilia on same, a shell, and other organs. This makes them quite big, something over .15 millimeters in diameter, and is why she can only accommodate a million at a time. A few days after the "discharge," as science unfeelingly puts it, she may be working as a male.

Now for the child and its labors. Not all the organs it grows for the larval period alone, to be cast off after fixing, are from the ancestral past. Those that are include the foot, which serves it for crawling, the teeth alongside the hinge of the shell, the anterior adductor muscle that the adults of the family gave up millions of years ago, and along with two other sets of sense organs, a little pair of eyes, no doubt as wise, if you could see them, as those of human babies. At the base of the foot is another very old invention called the byssus gland, which could be adapted for the manufacture of plastics and perhaps has been; it secretes a fluid that in contact with water turns into the thread used by mussels and some other bivalves for hanging on to rocks. This of course, and the organs of balance also in the foot, will be shed along with that extremity. Some of the work is for keeps; there are the beginnings of heart, kidney, gills, certain nerve ganglia, etc. Being itself plankton the creature feeds on plant cells about a hundred times smaller still.

All of which, wonderful as it is, is not much more so than our knowing it, thanks to a small number of people. There must be individuals like that among oysters too although brainless, geniuses on some organic level, going down our gullets or into the stomachs of starfish like any other fodder, or more likely dying in infancy. One in ten may survive the larval period, but the worst bridge to cross is settlement; only one in ten thousand manages it, genius or not.

The eyes seem to have a role in that; there is evidently an aversion to light. The larva swims with the velum upward and the foot protruded beside it, since it prefers an undersurface if one is available. It can change its mind, or whatever it has in place of one, a few times after feeling out a place, "looking for a home" as the weevil says in the song, a kind of choice and deliberation shared with many unrelated sea animals—a sporting provision, to give the race a fighting chance against odds that seem hopeless. But once it has really put its foot down it is through, for better or worse. Under the microscope it is seen rocking to and fro, presumably squeezing out from the gland in its foot the glue that will hold it. Then immediately it flops over to apply its so-called left or lower valve to the glue, which hardens in a few minutes, and that is that. One would suppose that now and then there would be a stupid little oyster in the litter that would stick by its right side, but there never is. A more curious fact is called "the effect of gregariousness," evidently a matter of chemical attraction. Like the barnacle and some other attached animals, the oyster tends to fix where others already have, another trait designed for the survival of the race, through greater chances of reproduction.

That, we must say, is a rather depressing item, and makes one realize how greatly the oyster's fascination is in its character as an individual, no matter how many millions of them there may be in any one spot. When we want a metaphor for unbearable crowds, or as one might put it nowadays, for the human situation, we think of flies, ants, lemmings, mussels, barnacles, but the oyster in our deeper thoughts is always singular. As it is in literature. Except in *Alice in Wonderland* where the author was dealing not with a personal image but a large commercial disaster ("And thick and fast they came at last,/ And more, and more, and more") the reference is always to the one, not the many. It is not "thick as oysters" but "silent as an oyster," "secretive as an oyster," "alone as an oyster," and that is why the first association of the word for most people is always "pearl." The error is only zoological; poetically, in a blabbing and currently swarming world, the thought is correct. Silence, privacy, the dignity of meeting one's fate without running squealing for help are the treasure the human mind can't stand to have relinquished altogether, so keeps at least as an image, and right on the tip of the tongue: say "oyster" in a crowded room and fifty voices will say "pearl" in unison, and the pearl that is really meant is being alone. And now we have to accommodate our image to an "effect of gregariousness"! It is quite disappointing; not that we are in any position to criticize.

The process of turning into an adult begins instantly, although it takes two days for the velum and three for the foot to vanish altogether, partly into the animal's own cannibalistic blood cells, or phagocytes. The back adductor muscle grows rapidly, while the anachronistic front one shrivels away. The suddenly enlarged mantle, with the function of creating shell, by the second day has made one that is not a mere extension of the late larval, or prodissoconch, shell but of quite different form.

The English word for the oyster at this stage, spat, is a miscarriage of language; really something should be done about it. Judging by Yonge's description, the whirl and commotion going on in the tiny organism are positively head-splitting—that's assuming proportions were otherwise and you could stick your head inside the shell a minute; and of course as in all biology not a part or motion, not a beat of one of the unnumerable cilia with their various functions, is unnecessary. Or if it is it had better be stopped; there is a lot too much happening that *is* necessary. Among other pieces of the devious and presumably flawless machinery is something common to all bivalves, called the crystalline style, "one of the most remarkable structures in the animal kingdom." Yonge adds in a footnote: "As I am reminded by Sir Julian Huxley, perhaps the only *rotating* part of any animal and the nearest approach to a wheel mechanism found in living organisms." It is a gelatinous rodlike affair with a head that goes around clockwise at the rate of sixty to seventy revolutions a minute in a certain area of the stomach, evidently with the combined roles of electric fan and blender, helping digestion and sending all food particles into contact with the walls of the stomach. Equally tireless, the cilia lining the stomach also help to keep everything in it in a constant state of upheaval, which sounds rather nauseating but suits the oyster. In fact the only part of the animal ever spoken of as being subject to fatigue is the pulling or "quick" muscle of the adductor, the part that wearies under the embrace of the starfish.

We are too tired contemplating the whole show to go on about the development of the gills, which work harder than all the rest of the anatomy put together. Shell growth seems a little less strenuous but that is probably an illusion. It takes place rather regularly in the infant, and in the adult by way of occasional sudden "shoots" which can push the margin out as far as a centimeter at a time, to be filled out and strengthened by more surreptitious work on the inside later. It is done mostly in spring and fall, not during the stress of procreation, and stops in winter. It is the succession of shoots that gives a very old shell of a *plate*, such as you rarely come across in the Morbihan but might pick up on the shore at Cancale, its beautiful layer on layer of petticoats on the outside, so strangely fragile, like ancient paper, at each scalloped edge, as against the nearly unbreakable pearly bulk within. But this is the grand old lady stage.

On the tiles, after the arduous work of metamorphosis, the oyster settles down to merely growing. Only the sexual organs are still to be added in time for a try-out, in the easier male phase first, the following year. In three or four weeks the shell will be clearly visible; by fall in a good year the little bumps may already be crowding each other at some spots on the tile.

The tourists leave precisely on September 1st.

EIGHT

On the day before Christmas, 1962, the region was struck by brutal cold. It would not have seemed particularly cold to Moscow or even New York. In that part of the world, where mimosa grows and everyone's living depends on a winter climate tempered to that extent by sea currents, it was alarming. The temperature swooped down to 18° F. and stayed there without relief until the 29th, aggravated by a bitter north wind that set in on the 26th. The workers of Locmariaquer, at the height of the season's business, found themselves without jobs. The best temperature for moving oysters is 48-50° F. At a few degrees below that, shipment becomes dangerous or impossible. However there is bound to be a crisis now and then in this occupation, and everyone naturally expected to go back to work in a few days. Instead, what started as a blow became a condition lasting three months. For most people there were no earnings again until late March, and by then the oysters were so stricken, rumor in

Paris had it that the "Belon" was finished forever; it was impossible that it should revive.

There had been abnormal cold and storms as early as November. Still, until Christmas the chantiers had all been bustling as usual. That really isn't the word for it. Inside the chantier buildings, most of them very small, there is a subdued racket of chatter and shells and wooden sabots on cement as the women go about the business of sorting, *le triage*, plus the noise of the sorting machine when the establishment is fancy enough to have one. But most of the work on the parks has a slow and lonely look. It is nothing like May and June. The work is of a different kind, and there is no whiteness around; the tiles with their wonderful accretion will mostly stay where they are until February. This is the time for moving the older crops, whether for shipment or relocation farther out from shore, and for the general work to be done off and on through the winter.

There are basins to scrub, crabs to kill, fences to mend, that is the park fences, fifty or more yards to a side and about eight inches high, with or without the out-thrust ledge and wire at the top, depending on the oysters' age and vulnerability. One of the few operations the oyster spares its keeper is feeding it, but he has to guard it from being fed on, and the wood too. Some oystermen import sacks of little snails from Ireland to eat algae. The parks have to be cleaned, and in some places spread with gravel to make the bottom firm and lessen the chance of smothering, which is most crucial for the yearlings and eighteen-months; for the older ones the danger is not so great. There is some piling up of silt at best and all the oysters have to be raked now and then, especially after a storm, either with hand rakes or at high tide with a harrow, about the width of a drag and pulled behind a boat in the same way. But the big event this time of year is change of domicile. Nearly all the millions of oysters are being moved, here and there by mechanical dragging and dumping but in great part by hand, and unless they are to be rebedded immediately, this involves a procedure called "fooling the oysters"—*tromper l'huître*.

It is actually a double procedure, to make the animal (1) clean itself out, and (2) learn to stay shut during transportation. Anthropologists would call this a rite of passage, and very solemn it is, not only as marking advancement to another stage of life, which the oyster may not appreciate, but because neglect of it can have, has had in the past, tragic consequences all round. The principals in the ceremony—youths, maidens and in-betweens but it goes for all later initiations too, through the last and most solemn, on the way to the table—go for three or four days into the stone or cement basins you have been seeing, called *dégorgeoirs*, where a judicious withholding of water and food at certain intervals causes them to disgorge their impurities. The educational part comes about as a result of the oyster's habit of "yawning" when exposed to air. The trickery, which must create some nervous breakdowns but leads to an extraordinary degree of self-control in the majority, lies in depriving it of water not just once but over and over when it is least expecting it. At the end of this trial it is conditioned not to yawn and thereby lose its liquid, and can travel for quite a long time without risk of suffocation.

It is an ancient practice, followed from way back in Brittany and the other great oyster regions. In the days before trucking and railroads, or when railroads were flourishing inland but didn't extend to many points on the coast, and the time of shipment by sailing vessels might be doubled by bad weather, few oysters would have lasted out a trip if they had not been cleansed and had their characters fortified this way. The army might learn a lot from it right now; so might we all. But among those who rushed into oystering in the big boom, at the end of the last century and beginning of this, there was the usual percentage of quick-money types who didn't bother to learn, and the consequent poisoning and deaths of consumers threatened for a time to ruin the whole industry. The worst offense was from Thau, the pond back of Sète on the Mediterranean, site of one of Coste's failures. Ignorant fishermen there, branching into oystering when it became profitable and usually lacking a basin or any kind of working establishment, would throw their oysters into any old mudhole on their way to market, and sometimes sold them straight from the beds with no interval at all. This was very bad for the oyster, which quickly gets sick when unable to rid itself of sediment and pseudo-faeces as its excretions are called, and killed a number of people. The damage was more extensive because the oysters in question were sold cheap and looked like a fine bargain, also because they came from deep water, so lacked even the routine training of most Morbihan oysters in staying shut. Other instances of poisoning and "contagious diseases," of what sort we haven't been able to discover, perhaps forms of hepatitis, were due simply to oysters being stored too long and sold rotten.

The trouble grew into a public scandal around 1906. A Dr. Chantemerre rallied a large section of the medical profession into denouncing the oyster as poisonous per se. Other doctors and scientists, better versed in the subject, supported the industry by pinpointing the abuse. There were some ugly debates, sales fell off drastically, and oystermen were at odds among themselves, some wanting to keep their own house in order while others argued for government measures to make the oyster respectable. That side won, after a good many years of struggle and of proposals that turned out to be unworkable. It was not until 1923 that the present system of government certification was finally put in force; the yards are inspected periodically and then are not given but are permitted to buy, by the thousand, the "tickets of salubrity" that must appear on every single box or basket of oysters headed for consumption. As most of the containers are quite small, containing no more than a few dozen *plates* between layers of seaweed, it makes for a lot of tickets. But that concerns only the "houses of expedition."

From Locmariaquer the bulk of the crop is leaving not for market but for the élevage whereby it will acquire all the fascinating names that appear on the menu—Whitstable, Colchester, Zeeland, Pied de Cheval, Belon and so on. The list is enormous and the names are far from meaningless, although the same parents will have produced some of all and statistically the chances are high, whether you are reading the menu in London, Amsterdam or Paris, that said parents were not more than ten miles from the bridge at La Trinité. The place and manner of raising, the chemistry of any particular water and the exact nature of food available to the growing oyster, make all the difference. At Paimpol or Morlaix, for instance, where some of the two-year-olds go for a year, the shell grows beautifully, making an oyster both handsome and resistant to enemies, but the body remains skimpy, so the next year should be spent somewhere else, as at Carantec, where there is more nourishment for the flesh and less for the shell.

That is a model itinerary and the one followed by the H. oysters, winding up with a final year of refinement at Riec-sur-Belon, which is not a famous old name for nothing. Whatever the diet is, that long winding tidal river-mouth is to the oyster what nearby Pont-Aven used to be to painting and is now for the gastronome, but that is misleading. The fat oyster is apt to be sick with liver trouble; its food should not be that rich. What is certainly not misleading is the feeling one has around there, and not only there, that it is logical for painters to do well where oysters do. If a marine biologist and an art critic could ever collaborate they might tell why, and the explanation might be quite simple, not esoteric at all—in some common equation of factors and atmospheres. It is a rare scene of oyster culture in France, even Thau where the oysters although no longer noxious are not very good, that doesn't wake the frustrated painter in every human breast. But the lights and patterns of the trade, shells, parks, tiles and tides, are only part of it, beautiful though they are. Even the least abstract of the Pont-Aven painters have done surprisingly little of that. The base of the equation would reach beyond these spectacles, into air, or chemistry.

For the oyster, and we'll drop the painter here though it's tempting not to, one requirement is not to live in the kind of muddy water that is likely to cause "chambering." There are various ways of becoming chambered—same in French, *chambrée*—a horrid condition although it doesn't need to affect the oyster as food, just so the spot is not punctured with a knife when the shell is opened. If it is, it lets out a revolting smell, sometimes of sulphuretted hydrogen from putrefied matter that may have been sealed in there for several years. This is quite a clever accomplishment on the oyster's part. The spot, on the inside of the shell and almost always in the cupped or left valve, can be caused by the worm Polydora, or by shrinkage of the body after spawning or after a change to saltier water; the mantle will then shrink, leaving a water pocket between it and the shell, and the oyster goes to work sealing it off with a wall of cut-rate chalky substance, good enough for the purpose and not nearly as hard to produce as the regular subnacreous shell lining. But the chamber may also be due to a speck of sand or other indigestible material that the oyster has not been able to expel, and that kind of chambering is commoner on some grounds than others. There is a good deal of it in full-grown oysters from the Gulf of the Morbihan, and relatively little at Riec.

But Riec is only one destination out of hundreds. The oyster babies of Locmariaquer are being scattered over their full adult range from the Mediterranean to Norway, many to England, a few in recent years to Canada and Maine, where efforts have been made to establish the species, and within a year many siblings that have been separated would not recognize one another. Many humans too will be speaking of them as if they were of different breeds, and will think that they are. Even leaving aside the crucial differences in taste, it is hard to imagine the huge creature called the Horse's Hoof, from Cancale, being of the same stock and womb as some of the beautiful flat smooth-shelled varieties found in England, or those bumpy others hardly distinguishable by sight from Portuguese though very much so to the palate, from certain rough inlets not far from Cancale, on the Channel coast of Brittany.

Not far, but a world away to the oyster, and sometimes to the oysterman too. There are places up there, around Tréguier for one, looking out to the fearful legend-laden rocks that Chaucer knew well, or under the crags where the English castles used to perch, that seem to offer so hard a life to the oysterman, you have to marvel at his keeping at it past the Middle Ages. Habit or poverty would not explain it; it would have to be a love affair, and so in some version it is. There is one oysterman in that region, one of the few not born to it, who used to be a rather well-off chemist in Paris. He gave it up for the happiness of pushing out over his parks in a rowboat through winter storms, for in ten years he hasn't made enough to have any helpers or machinery, while his wife when the pickings are thin stands hawking their wares and blowing her hands for cold on a street-corner in Tréguier, as the women of Arcachon did a hundred years ago in Bordeaux. You couldn't just like it, it would have to be love and perhaps with a sea-shelf of legend at the roots, perhaps the business would die out if at Christmas you couldn't hear the bell ringing for mass at the bottom of the Great Rock; these are only perhaps's, perhaps of some economic importance. The ex-chemist and his wife are happy; have never spoken of turning back; are among those from Tréguier and thereabouts, Saint Yves's district and Du Guesclin's, who turn up in Locmariaquer in their small trucks every fall to take home their yearly stock of eighteen-months. Not naissain in that particular case; their patch of water is too wild.

Other trucks are there from mild Marennes, Marennes the green, only what greens the oyster is described at least in French parlance as blue—*la Navicule bleue*, a diatom. The same effect is produced in certain English creeks, and is and was so prized that in the 18th Century the English oyster-eater was sometimes given a ghastly imitation of it, in oysters that had been thrown for a few days into pits full of green scum. Another bad green, rather rare, comes from excessive copper, which gives the oyster a disease called leucocytosis. In those cases the whole body is discolored. The good green, gourmet's delight, is only in the gills, the rest of the body remaining white. Like most good things the process was discovered by accident, and the name is missing in history of the "bold man" who first bit into such an oyster; or he may have been stupid; or perhaps some diabolical son-in-law gave a plateful to the old lady and was paid off by having her live to be a hundred and demanding them every day. In some such empirical fashion the green oyster, which does have health-giving properties beyond the normal, came in, but in a sense Mme. de Maintenon was right to be skeptical, whether her motives were honest or not, for nobody could have told her with certainty, then or for three hundred years after, what it was that produced the phenomenon; it might have been something nasty.

It did make for some nasty controversy in learned circles, nearly bringing on a French-Italian war around the turn of the century. The first work on this problem was done in the 1820's and then dropped until the 60's, when it was taken up by several scientists including Coste. He worked and speculated on it at length, analyzing the water and taking note of all observable facts, as that the greening stopped during the season of procreation, resuming in August. Of the several theories held at the time, involving minerals in the water, liver disease, an "animalcule," etc., he finally leaned toward the one that took the nature of the soil, through some undetermined component, to be the prime source. He was wrong; not totally, since the *Navicule bleue* (Navicula fusiformis or Navicula ostrearia), identified soon after as the greening agent in this connection, does depend for that work on certain micro-organisms in the soil. But it is not a question of a direct effect of minerals, such as give a distinct taste and tint, brown, bluish or slightly orange, to oysters in many other places. Our friend Bashford Dean, whom we left standing by the now-vanished lake in Locmariaquer, contributed to the further research in the matter, along with a distinguished company of brains in France and England, and by 1890 the main fact of the case was generally accepted. One of the Italians who still held out against it was accused by a Frenchman of having studied not the true "huître de Marennes," but the rotten "huître de Venise"; verbally what was involved was more like rotten eggs. This great battle, which took place as late as 1909, is known as the Sauvageau-Toni exchange and there are people alive today in several countries who remember it with horror. The *Navicule bleue* side won.

An odd fact of modern human life comes in here. Whoever invented the little plastic dinosaurs to be found in every toyshop nowadays had the right idea; dinosaurs are for children. For the contemporary grown-up mind, insofar as we continue to have such an organ, jaded and punch-drunk from every sort of bigness from advertising on out, only time and the smallest forms of life are still spacious. Here the imagination can still stretch, excitement is still possible, the Truth, or whatever bit of it one arrives at, seems still to retain its sacred majesty. The proof of this proposition is that nothing in oyster literature is as fascinating as Dr. Ranson's description of the minute vegetable that creates a business worth billions of francs in Marennes. At least at the moment it is considered a vegetable, of the general family of Algae, in spite of a capacity to move under its own power; it may find itself in some other category a decade or two from now. Anyway it lives, it moves, it is 62 to 130 microns (thousandths of a millimeter) long by 5 to 12 microns wide, which means you could put several thousand into the shell of an oyster larva, itself invisible; in a good season it covers the mud-walled *claires* of Marennes with a solid green mat over a total of several square kilometers and with a total weight estimated in tons; and it fooled all the scientists for years by having a double personality as dramatic as any in the annals of psychoanalysis.

There are two kinds of marine diatoms, those hugging the coastal bottoms, called benthonic, and those of quite different habits and habitats, which travel far and wide in the currents and are called planktonic. Our particular Navicula was originally spotted in the latter class and its protoplasm was found to be colorless, so of course nobody thought of connecting it with something that was obviously both benthonic and green. The sleuth in the last stage of the case, if one could ever speak of a last stage in these matters, was Ranson. He discovered that given proper food and other conditions the little speck could acquire color and completely change its way of life, and furthermore that the extraordinary transformation was in large part due to the mucus, or more precisely to the form of sugar called glucosamine contained in the mucus, excreted by oysters. This mucus, which is being created and poured out in vast quantities wherever there are large numbers of oysters, is what gives the *vase* its gluish consistency, so unlike any ordinary sea mud. Mussels and other lamellibranchs change their environment and spoil the swimming for people in the same way, that is it feels the same to a human foot, but not to the *Navicule bleue*, which in the presence of mussel mucus remains resolutely planktonic. Only the oyster can woo it into settling down and turning blue, and only in a little area around Marennes are other conditions right on a big enough scale or enough of the time for an industry to be based on it. The fact that other molluscs if bedded there will also green is evidently due to the presence of oysters. We find it very hard to think of this enthralling little creature as a vegetable, but that is probably because the word has been so maligned. We will never use it in derogation again, and have been happy to learn that this member of the kingdom is not mortally sick when it takes up with the oyster, a possibility that was considered for a time. It is perfectly all right, in fact healthier and more energetic than ever, and can go back to being planktonic any time it wants.

The *claires* in which all this goes on have no resemblance to oystering arrangements anywhere else. They come under water only in the spring tides and with their mud embankments look more like rice paddies; some are not even by the shore but back in the fields where they are irrigated from little tidal canals. It is perhaps not much less lovely as a scene than it was a hundred years ago, but the work is hastier. In Coste's time the best-quality oyster had had the special treatment for at least a year, usually longer, and he deplored the quick greening for no more than three or four weeks, that some oystermen were beginning to practice; it was a cheat, he wrote, coloring the gills without any real change to the system and therefore the taste and properties of the oyster. Whether he was right or not, two or three weeks in a *claire* is the norm now. Out beyond you see the parks of the usual sort, for ordinary élevage, and at a few points near sufficient currents there are masses of collectors, mostly of the rosary type, strung with pieces of slate or shell. In general the district has to import a large proportion of the seed it needs, but there are oystermen who rear only their own, both *plate* and Portuguese. One of these, a sad, thin, swarthy man in his fifties, slowly paces his dykes in the off season, seeing to this and that; this is at Bourcefranc, a few miles from Marennes. The building is quiet, a few oysters are quietly turning green; nothing much. It depresses him; in the last year he has suddenly felt tired. For generations his family did a small business there, which by an extreme of hard work and deprivation he turned into a big one; he does everything, all the stages, from collecting through *l'expédition*, and his son wants to be a singer in Paris. A singer! Could you get farther from oysters than that? The work is too much now, and there is no point to it; he doesn't know why he goes on.

He doesn't know what turns the oyster green either. He has heard theories, and heard tell of the *Navicule bleue*, but hasn't looked into it although to do so he would only have to walk fifty yards. One of the tiniest and most delightful museums in France is there almost at his door, the Musée de l'Huître, gotten up a few years ago by the private initiative of a group of oystermen, with a little government help at the end. The group was evidently of a different intellectual cut from our weary friend, or perhaps something about the way it was done piqued him and made it beneath his dignity to learn from it. Aside from Ranson on the Navicula, which is all there with pictures, there are real oyster fossils the size of an elephant's ear—well, elephant's hoof, at least in diameter, and diagrams of larvae and all ages of spat and types of collectors and just about everything else you would be craving to know about if you hadn't read a book instead, and in the plainest possible little shed at one end of the building with three rough tables in it you can, most marvelous to tell and to remember, *eat oysters.* Not Portuguese although you could; you will save money on something else.

And all your facts and knowings slip away, under the poignant stab of a taste you had thought no food, perhaps no anything, could give again. Nobody knows from how far the breakers roll in.

THE sorting machine, a simple and noisy thing, works by the kind of motion that used to be called the shimmies. From a metal bin high up at one end, holding several bushels, the oysters are joggled down along a sloping chute with a series of round and progressively bigger holes in it. Sand and gravel are shaken out at the top, then the smaller oysters, and so on down. Not that oysters of different ages are brought in mixed as they would be from natural beds; the usual practice is to deal with the eighteen-months first and the two-year-olds later, but some do better than others in any group and the range in size is considerable. The room, which generally means the whole small interior of the building, is of course unheated. The women sit on stools or benches along both sides of the chute, taking out the baskets from underneath as they are filled and picking over the crop by hand. Far more commonly in the Morbihan the whole procedure is by hand and the difference in time is not great. There is too much the machine can't do, such as weed out the misshapen oysters, and with the *plate*, in contrast to the Portuguese, shape is of the utmost concern. It is a question of noblesse oblige. O. edulis has duties as well as privileges and not the least of them is to look well on the plate.

Then washing, packing, shipping. The total bulk handled by the yards is never the same two years running; it depends on all the variables, of rainfall, temperature, pests. At certain points in the region, the normal weight per thousand two-year-olds is 25 to 30 kilos, but right off Locmariaquer that is average for oysters a year older, hence its specializing in the very young. In 1961 the figure per thousand two-year-olds, in the better places for that stage, was down to 18 kilos and nobody seemed to be bothering much why; that much fall-off is to be expected now and then, as in quality of wines. It was nothing like the dramatic decline, not in size but numbers, going on for some years now only twenty miles away in the River of Étel. The devil had been brooding all these centuries over what Saint Cado did to him on his bridge, tricking him with a mere cat, so in 1956 he came back in the form of a fish and so far the saint hasn't done anything about it; perhaps he doesn't like the way his chapel has been restored, for which you can't blame him but it is hard on the oystermen. The fish, called the *dorade royale*, commonly an habitué of more southern waters, has jaws like a tiger's and has been arriving every year as soon as the water warms up to 60° F., to spend the summer eating oysters. One crack of the teeth and that's it, and back for more with every tide all summer long. They say, as people have always said of the devil, that he might disappear again the way he came, but there are oystermen along the river who may be ruined first.

As for the economic value of the total crop, in the Morbihan or in France, that is anybody's guess; which is to say that any "reputable source" would laugh at you for asking. There are probably figures on it somewhere in Paris, pulled out of a hat by some harried under-official whose chief required them. An accurate report of income by a person in this business, as no doubt in some others in France, would not raise his social standing in the community, and might cause him to be carried off to an institution.

Back to the chantiers there come at last, assuming a normal course of events, the naissain. It is the spring procedure in reverse, but without the regatta, the fun; there is no one precise time for setting out, and mimosa notwithstanding, the water off any part of Brittany in February and March and the whipping rains are not conducive to cheer. Getting the bouquets back onto the barge is something to endure and get over with, no matter how lucky the catch, but it can't be hurried; the cargo is fragile as porcelain and to be put down tenderly, whether in thunder and lightning or under a pale boon of sun. Then back home at the yard the muck is hosed off and the most delicate job of all, *le détroquage*, begins. Venetian glass or cloisonné work of the old days that used to require the eyes and fingers of young children are what you think of, and yet a woman, with the cold and swollen hands of this calling, "if she works well" as they say, can remove ten kilos of the tiny and still nearly translucent shells from the tiles in a day, not counting mortalities which are many. In numbers, this is somewhere in the neighborhood of three thousand living pieces, which stand a good chance of being killed by the slightest clumsiness or slip of a knife. But the naissain are too small to count, and are sold by the ton, not by number like the bigger oysters. They come off the tile with a little spot of chalk on the left valve, at the point of attachment, and keep it for the rest of their lives. It is called the "heel" of the oyster, and distinguishes it from those dragged from natural beds.

Many are sold in the neighborhood, the big producers buying from the small, at a price set beforehand and which brings bitter complaints at times from the family-scale operators. They have little say in the matter, and often having no parks or not enough to keep the seed until the next year, are obliged to sell whatever the price.

That was one grievance they were spared in 1963. By February nearly all the naissain were dead.

Only one person in Locmariaquer had known such a winter before; few had anywhere in Europe. Halfway down the coast of Italy there were days when fish were washed up on shore by the thousands, dead and dying of cold. In Holland 98 per cent of the oysters were killed. In the Auray region, what work there was for anyone around the parks was mostly clearing away wreckage when that was possible, and corpses of oysters, for the fish that would normally have eaten the dead had frozen too. But the worst damage was not known until six weeks after the worst of the cold.

To compound the trials of fish and men, the wind continued from the north and northeast through a great part of the bitter season. The Postmistress, rheum in her eyes and the outermost plane of her distinguished nose red as a feather, sang and sang against the banging of shutters and shriek of wind in the roof-gutters, "Mais oui," "Mais non," "Mais certainement . . ." Of course it was nothing like summer, with the mobs of foreigners in there every minute demanding mail, leaving forwarding addresses, telephoning their relatives all over France, usually just to say they were having the time of their lives, or the contrary. But for winter it was something new. After the first cold wave around Christmas there had been a slight relief, very slight but enough to give people hope; the oysters in the parks had been known to survive that degree of cold, over short periods. The real blow came on January 11th. The mercury went to a new low and from that to worse, smashing all records, and that time the wave lasted with no break worth mentioning until February 5th. Even then, for some five weeks more, the relief was negligible, to people in general and the Postmistress in particular, for along with the Mayor's office hers was one of the last to be affected by the *crise du charbon*. People who had never used a postage stamp in their lives took to stopping in to get warm, before heading back across the frozen fields with the bread or whatever they still had the money to buy, which wasn't much by that time.

The other business had already begun in the middle of January. That was making out the withdrawal forms of the National Box for Savings, and in certain cases, knowing it would all go for drink by the next day, she was hard put to it not to be just a trifle *irritée*. She expressed herself in a dignified if somewhat startling attitude of levitation, in which for a second every part of her seemed slowly to rise and part from its natural support, neck from shoulders, feet from floor, eyebrows from the usual place like two pins on a magnet, while in still more superb detachment there issued from somewhere the well-known aria ending, "Et ici la signature . . ." But for most it was not like that and from her bureaucratic security she suffered, hearing over and over the same explanation, along with the new crises of each separate day, until it came about that she was not always in command of her repertoire or vice versa. She then might be heard to utter, in the speaking voice of plain sympathy, "Oui, c'est dur," or no less genuinely, as on January 17th and 18th, the pale eyebrows soaring dangerously roofwards but still not leaving her face, since its other Gothic lines and pieces managed to fly into some sort of accommodation with them, "Pas possible!"

Those were the two days when the temperature went lowest, and it was a period of *mortes eaux*—the neap tide. The ominous words "la glace" were in everyone's mouth. The upper stretches of all the estuaries, of Crach, Auray, La Trinité, were frozen over; the parks and basins froze solid; a band of ice eight or nine inches thick formed around the whole Bay of Quiberon. In the harbor of La Trinité alone, where the narrowing channel on both sides of the big bridge finally vanished overnight, several million tiles with the year's naissain on them were locked in ice. Ten days later, although the air was not much warmer, the region echoed with the crash and boom of ice breaking in the spring tide. Oysters trapped in the chunks were swept out to sea; others were driven under the *vase* by the weight of the ice, or by storms. The careful arrangements of bouquets were bashed down like matchsticks and on many of those that stayed in place the chalk peeled or was knocked off and the seed went with the chips. The basins stayed frozen much longer.

There was rain for a couple of days, and a faint resurgence of hope, followed by snow.

Yvette was not one of those who went to the Post Office for their savings; she had none. In the fall when the children had been given a medical examination at school her little boy Yves had been found to have TB; nothing serious, they said, only he must go to the sanitarium for a year. The Family Assistance paid most of it but it was thirty miles away and she had spent her summer's earnings on the trips once a month to see him, besides medicines and extras. Just the same, she was glad now; he was going to school there and was happy, and that was one place that would get coal, as long as there was any at all. So really, although there was a great ache in her of missing him all the time, especially in February when the bus fare was out of the question, the little spot on his lung was a blessing. Another was the Janec farm; they wouldn't starve after all and there was heat from the cows; she could sleep there in a pinch, not with Jean-François though. Alice, ten years old now, all jokes and fun and quite fond of her father in the daytime, still screamed horribly if he was under the same roof at night. For her it was always that same night, the time he cut Yvette's head open. They even had some family parties in the cow barn that winter, with Léa and her husband and Yann and sometimes the sisters from Auray, with plenty of applejack and funny stories, there were so many times when there was nothing else to do. Of course it was cold there too. Still the really bad things were only Jean-François and the sea, and the gloomy talk you had to hear, people saying the oysters would never come back in Locmariaquer and what would they all do.

The sea was terrible. At the next low tide after the big freeze she and Léa went out on the still partly frozen bay by Françoise's hut, with knives and baskets, just as *tout le monde* was doing, there and in every other cove and inlet, looking for something to sell. Nearly all the oyster-workers from that side of town were there—Mme. Aurogné for one with poor lumbering Jean-Pierre who never could catch anything but she hadn't dared leave him at home; lately he had been having alternating fits of semi-coma and violence, evidently from the cold, and she wanted to keep him moving around. As it turned out he got as much as anybody else. The palourdes were all empty shells; nearly everything was dead, and when the tide began driving the farthest figures in, it was nothing like the serene loneliness of the summer but a huddle of cold and dejection. Nevertheless they were out trying again the next day and the next, and here and there a live crab or a few snails were found. The most desperate was Françoise, a blanket pinned around her shoulders, down on her knees poking under the frozen seaweed and around the stones. Her mother, so the doctor and Sister Sophie had just told her, had pneumonia and was past moving to the clinic in Auray even if she had consented to go, which she would not. She was ninety-seven after all and couldn't possibly live beyond the week but Françoise was seized by an obsession of hope. The tidbits she brought from the sea had done miracles so long for that ancient body and spirit, she would not believe they were beyond the same miracle now. She thought that if she could find even a sea slug, anything alive, to tempt her, the fever would start going down. So she lingered, hobbling out farther toward the open sea after each try even after the others had turned back, thrusting her knife more and more wildly into the icy hummocks. Yvette and Léa went after her finally and gave her the three small mussels they had found, which were dead but still edible; her sabots were full of water and they had almost to carry her to the house.

The little fire she had left had gone out, and the wheezing old bundle under the quilts, on the bed where Marie-Yvette had screamed the last of her life away, was burning with fever worse than before. They couldn't get the mussels down her throat; an awful roiling and rejection from the depths, as in some deep rock-pool on the Côte Sauvage under the incoming tide, set in when they tried, spewing the slippery morsels out into the hot folds of her neck and under the covers. So Yvette and her sister left, saying they would bring milk and firewood in the evening and would send someone for Sister Sophie again.

Jean-François was fairly sober that week, and working. In fact he was feeling rather heroic. The H. parks were the most exposed in the area and the chestnut fences had been carried away in the ice. Now after the break-up ice was forming again, so he and Yann and Poujade were out keeping fires going around the parks for as long as they were uncovered, to try to save whatever was left. It was impossible to tell how much that might be. Two-thirds of the oysters in the parks had vanished altogether and there were empty shells everywhere, as many beyond the barriers as within, but there were some that might pull through. They were burning everything possible, first the piles of discarded worm-eaten pickets, then old sawhorses and any lumber that could be dispensed with from the chantier. Monsieur H. himself kept the wood coming and tore down a shed for more. He was looking half maniacal from fatigue in those days, spending few hours in sleep and half his waking ones in a race over the glazed and twisting roads to take up the struggle in his other places; however he kept his wits in order and his voice calm, better than some of his friends in the business. Actually, with the exception of Cancale, the north shore was suffering least but that couldn't be counted on at the time. Madame too had turned into a blazing demon of determination. The lovely négligée in which she had been wont to stretch her catlike limbs in the period between bed and convertible was forgotten; so were the children, a little more than before; twice, as she sped with new orders to Paimpol and Riec, her car turned completely around on the ice and she went on without letting the accelerator off the floor. "It's Jeanne la Flamme," said Poujade with a wry touch of admiration as they warmed their hands at one of their fires, "descended among us. Only she can't burn Frenchmen any more. C'est pas permis." The others laughed a little, glad of anything to take their minds off the wind. Yann was as haggard as the boss, tormented by his responsibilities as foreman in the face of such impotence; they were pouring water in a sieve but he would get up in the night to put in another drop, take some new measure.

Just then they were about to move a trayful of eighteen-months, pried with huge effort from the frozen waste, to one of the parks farthest out where the damage was less. "Idiot!" "Imbécile!" "Nom de Dieu! . . ." Poujade, of course, being the public servant again in the one way he really could. A gust of smoke had sent him skipping backwards onto the *caisse* and three or four hundred oysters were now to be gathered up all over again. "Well, now, if the state could be put to rights as easily as that . . ." He was uncertain whether to make it a whimper or a declamation and besides his teeth were chattering, so the response to his mumble was only a kick on the behind from Jean-François. There were a few other fires visible far away, for the same purpose as theirs, and more hidden by the ragged loops of the shoreline. The centuries had slipped; it was the winter scene of the times of Blois and Montfort, only the fires were up on shore then and Jean-François could have done great deeds every day. However the present challenge was better than none and he was full of theories about it, depending on what he had had to drink: it was all from the atomic bomb; oysters didn't freeze, they only asphyxiated, etc., always ending with the happy declaration that the savants didn't know any more than he did. The woman who did house-cleaning at the Institut des Pêches had told him so.

He was right to this extent, that the Institut, along with the other oyster-savants of Europe, was having the chance of the century to learn about the effects of cold on that form of life, and was keeping records as never before. Something had been known, as that cold weakens the adductor muscle, causing the oyster to "yawn," and that at a temperature below 40° F. the action of the cilia practically stops. What was new for scientific purposes was the massiveness of the evidence and the duration of the cold. That, and not any one freeze, appeared when it was all over to have been the main factor. The oysters, which had gone on eating until the middle of January, then stopped and ate nothing more until early March, although the diatoms making up their normal diet were extraordinarily abundant. In short they went into an unnatural degree of hibernation; their physiological functions came to a near stop. This saved the stock in basins from drowning on the excess of oxygen under the ice, when their need for oxygen was at a minimum, but also made it impossible for them to get rid of sediment when the bottoms were stirred up by storms or harrowing. On the other hand it seemed the harrowing must be done when possible, to prevent total burial.

Another manner of death for the oyster, according to speculation in some quarters, might be by the formation of ice crystals in the body cells, which under any ensuing blow or movement would tear and bruise the tissues. But Jean-François's report from the Institut des Pêches didn't extend to that, nor to the parasite that assailed the weakened oysters and in the final reckoning was found to have been the main cause of death. The bureau itself didn't have the full data on that until much later. In February the final reckoning was still far off, except for the usual prophets of doom, who seemed to have a good deal more than usual on their side.

These considerations, although in a general way he was aware of how deeply his life was entangled with them, were not uppermost in Monsieur Giannot's mind that evening when the child, Yvette's daughter, came in. He had troubles enough as a farmer without worrying about the oysters too that winter. He was milking the cows and hearing his son's lessons, grammar, history and arithmetic, as he did on every schoolday at milking time; he was secretly proud of Morvan, who was strong and brave and had just turned seven, and was determined that he should show the Abbé his best. Beyond that, to what the boy might do or become some day, it was not easy for the father to think, but at least this much was clear, that he was not to be slack in learning or in other duties. Since he was four he had been bringing the cows home by himself, and he had never been afraid as some other children would be when the wild one's hobble broke or the path went under in the big tides.

Tonight he had tried to beg off and Giannot had spoken harshly to him and pushed him from the kitchen into where the cows were, with his books. Now he stood under the bare barn light bulb, reciting in a pinched unnatural voice against the ping and sizzle of milk going into the pail, "On March 26, 1351, between Josselin and Ploërmel, there took place the Combat of the Thirty. The English were armed with lances and at first the Bretons had the worst of it . . ." The voice trailed off. "In whose reign was this combat?" demanded Giannot, moving to the next cow. He went on more crossly, trying to help Morvan keep his tears back because he knew he would be ashamed later to have cried over the cold, over being cold himself and his mother being sick in bed from it in the room beyond the kitchen where they all slept. "A king? a duke? you haven't said what they were fighting about to begin with. Come, the udders are warm, we'll pull together." So in a minute the boy was laughing at having his father's tremendous arm around him as they balanced on the little stool together, and was just getting to where the knight said, "Drink your blood, Beaumanoir, and your thirst will pass"—his favorite part, when Alice ran in. "We have to go for Sister Sophie," she said to Morvan. "Pour la maman de Françoise. Tu viens?" He looked at his father, asking, and Giannot looked at the door and the wind. "It's almost dark . . ." But he knew he would have to nod, there was nobody else free to go. "Leave the bicycles there, at least. She can bring you back. And watch out for the ice." It seemed to him that was the only word he had heard for weeks, *la glace*, and it had come to encase everything. His wife was young and strong, she would be well in a few days, but the ice that was over the fields and the sea was on her heart too, and now suddenly it was all he could see on the years that lay waiting for Morvan, He had been wrong to name him for so great a hero, even if the name had always been in the family; it came too near to suiting him. What could there be for such a boy? Ten cows were milked, leaving twelve to do. He moved on again, and was filled with a great sadness for the old woman dying over there by the frozen shore, whether because she had lived nearly a hundred years or must die in such cruel weather he couldn't tell.

It was exciting as far as the oak grove by the pink mansion. In disturbed reflection the pale half-moon throbbed on the platter of ice below the dyke, where last summer the four images of horses had played among their private clouds. The children's bicycle tires tinkled pleasantly on the crystals along the dirt road, beaten up by cows and Giannot's tractor. Then all at once passing the dark grove to go up on the dyke the wind shot out at them and snatched the moon away. The tide was up, grinding at the old oyster basins, fighting the ice in front of the dyke and the haunt, whatever it was, in the huge empty house. They got off their bikes to cross. "Ça fait peur," Alice said. She didn't really mind it so much there in the open, except for her nose being about to crack off. It was the other side she was looking at, to the start of the two miles between the black hedges that were so high over their heads and in certain places would touch them on both sides. Morvan took her hand. "Never mind . . ." and then he said a wonderful thing, that made all the difference: "Il faut, tu sais." Yes, she knew that too; *they must.* "Anyway," he shouted back more lightheartedly as they got on their bikes again, "it's just like every day. The big ones don't know the path the way we do."

It was at just about that time that a great piece of the Côte Sauvage, a chunk of overhanging rock big as a farmhouse, edged by a crevice that ice had been pushing at for two weeks, broke off at last under the furious reach of the waves at high tide and fell some two hundred feet into the sea.

The old woman rallied remarkably when Sister Sophie was there. It was the first time in two days that she had been able to say real words. It was not too cold in the hut; the nuns had given up one of their small gas heaters and had it put in there a little earlier. Françoise had gathered up the three mussels and her mother was able to get them down this time, and said they were good. Her forehead, damp and ridged with age like the floor of the sea, became almost cool, just as Françoise had predicted. In her last lucid time earlier in the week she had said, "My poor daughter, who will take care of you when you are old?" and Françoise, who didn't care to remind her that she was seventy, dreaded hearing it again. But it seemed it was something else she was trying to say now, and Sister Sophie who didn't understand much Breton thought she was getting it wrong. She wasn't; bending close, they both made out the words. She was speaking of the cold, and the oysters dying, and the defeat of the Emperor, and when her glance sought the window they could tell that she was searching the wall of the great farmhouse that belonged to the Parisians now. "Ah . . ." breathed Françoise, in terror, but that was only for a few seconds. It seemed peaceful and proper to have moved into that other terrible winter and into their own house, with her mother a little girl of five and herself the shapeless spirit, in the company of all the unborn and the dead. "Yes, Maman, the Emperor has been taken prisoner." There was a gurgle, of some new discomfort perhaps, as the great bulk resettled itself one last time. "Good," she said distinctly. "Good news for Bretons."

For Yann standing at the end of that bed at midnight the waves at his back were not audible, nor the Abbé's voice intoning the Last Sacrament, over the screams of ten years ago. He hadn't broken down then, but couldn't stand it now and had to go out before the end. Around Yvette and Mme. Janec and the others crowded into the little room the walls grew higher and moved slowly out; the Abbé's voice was doing it, and he himself was strangely tall, unless it was that the old woman with her last breaths was really exercising such power and they were in the farmhouse after all. It was not positively beyond belief, knowing her as they did. But then the walls were higher than any walls and were not even of granite any more, and across the ravaged old features of Françoise, etched beyond any possibility of further suffering, there was something like a movement of whiteness, emanating from the Abbé and from four horses that were no longer in the field.

THE parasite, called Hexamita, hardly ever found in oysters from the parks in normal times, took over some 80 per cent of them that winter. As late as April many were still dying of it, and none of the moribund oysters that were examined throughout the winter was without it. Among survivors there was a very high rate of chambering; it had been a rough time in every way and they had been too weak and comatose to get rid of extraneous matter. Those were the lucky ones. Freezing broke down the chalky walls of the chambers, and unless there was a strong subnacreous lining, the deposit affected the adductor muscle, and the oyster yawned and died. In general, in the Morbihan as elsewhere, the mortality was by far the greatest in the basins and the parks closest to shore; the two- and three-year-olds farther out and so uncovering less often did better on the whole, although not in places where the bottom, sand or mud, was excessively churned up. The Portuguese, as expected, came through better than O. edulis, with the same exception; at Pénerf more were buried than if they had been *plates*, since they were on softer ground. The loss of naissain on the bouquets was always greatest inshore and least out by the channel, and greater on the bouquets forming the outside border of a block than on the inside.

There were various estimates of the losses, but it was hardly worth making one for Locmariaquer. There, for all practical purposes, there were no naissain left at all and almost no eighteen-months, making a shortage that must extend over the next three years even if there were no further trouble. Nevertheless, and before it was time to start scraping the tiles again, there was work at some of the chantiers. Like a great wounded fish the community moved slowly, a little wild-eyed, back to its rightful ways. There were some differences from 1870; there was a good deal of insurance to be claimed and the government eased up on credit to the industry. The hope for the future was in the natural banks that the Institut des Pêches had reconstituted, against strong opposition and conflicting interests, over the last twenty years; also in the fact that the adults of the breed had already been moved in large part to the north coast, which by a freakish indulgence of fate and climate was not badly hit. Meanwhile a few batches of oysters were dragged from deep-water beds, to keep things going.

Those were what Yvette was working on the first week in May, when the gypsies set up their camp again opposite the Table des Marchands. Even lying broken on the ground as it is in this century, instead of in one piece on the rakish incredible perch it maintained for three or four thousand years before, the great dolmen must speak to them in some way, about the sun and other untopical matters, for them always to choose that spot. Yvette liked going by it too, at least in the daytime, but then everything was making her feel good that week. The sun was shining, her flowers and vegetables were coming up, the predictions were very good for next year's naissain, in spite of everything, and Alice was being such a little comedian now that it was warm again, she had everyone laughing. The two of them had been singing funny songs all the way from school, and perhaps that was why the young gypsy woman with all the children, the same one who had been there often before, smiled at her for the first time. Of course Yvette didn't think much of gypsies but she had an impulse to turn off the motor of the bike for a minute and ask how the winter had been for them. The woman shrugged and said, "You are the most beautiful of this countryside." Yvette blushed and thanked her, and grumbled to Alice when they had moved on, "Oh, the flattery; just like them. Now what will she be trying to steal?" Alice who was on the little extra seat behind gave her a tremendous squeeze, and said without being funny at all, "But it's true, Maman, you *are* the most beautiful, in all Locmariaquer." She ended more grandly, "In all France."

BENEFICENT Oyster, good to taste, good for the stomach and the soul, grant us the blessing of your further mystery.

About the author

About the book

Read on

Insights,
Interviews
& More . . .

*

Eleanor Clark and James Agee, 1949 (Photograph by Walker Evans. Reproduced by permission of the Metropolitan Museum of Art, the Walker Evans Archive, 1994 [1994.253.668.1–3]. Copyright © by the Walker Evans Archive, the Metropolitan Museum of Art.)

About the author

Meet Eleanor Clark

SPEND A SHORT WHILE READING about Eleanor Clark—whose name persists in the letters and biographies of John Cheever, Katherine Anne Porter, and other knights of the plume—and you'll again and again come into collision with the word "ferocious": Eleanor the "ferocious driver"; Eleanor the "ferocious reader"; Eleanor the "ferocious skier"; Eleanor the "ferocious debater." Certainly she approached life with a ferocious hunger, whether exploring Italy (*Rome and a Villa,* 1952) or, many years later, combating the eye disease known as macular degeneration (*Eyes, Etc.: A Memoir,* 1977).

Eleanor Clark was born in 1913 and raised on a Connecticut farm. She attended a one-room schoolhouse, to which she walked with sister Eunice, two years her senior. The girls shared an unwearying interest in languages, having spent two years "at a convent school on the French-Italian border," according to her niece, Rebecca Jessup. "They were at ease speaking French, Italian, or Latin. My mother remembered reading Caesar in Latin under the covers, like something forbidden."

The sisters came into their intellectual exuberance courtesy of their mother, Vassar graduate Eleanor Phelps Clark. "She was a threadbare but zealously educated New Englander," says Eleanor's daughter, Rosanna Warren. "She had done graduate work in French and Italian but never got her PhD." (Eleanor Phelps Clark divorced her husband, Frederick Huntington Clark, when the girls were young.)

A young Eleanor Clark (Photograph courtesy of Rosanna Warren and Gabriel Warren)

Both Eleanor and Eunice attended Vassar, where they founded a radical magazine with Mary McCarthy and Elizabeth Bishop. The young women "could never be sure who had contributed what when they met at a Poughkeepsie speakeasy to lay out an issue" (Carol Brightman, *Writing Dangerously*).

"Eleanor possessed the glamour, gray matter, and ambition to compete with McCarthy," novelist Michael Mewshaw has observed, and the writers did in fact remain slight nemeses throughout their lives. Said Eleanor of McCarthy: "She was highly competitive. When I was doing *The Oysters of Locmariaquer* . . . I said to somebody, 'Don't tell Mary, she'll do a book on clams' " *(The American Prospect)*.

Upon graduating from Vassar, Eleanor lived on the cheap in the West ▸

Meet Eleanor Clark ***(continued)***

Village and contributed articles to magazines. "[O]ne day she got up the nerve to enter the office of *The New Republic* and ask Malcom Cowley for a book to review," wrote Joseph Blotner in *Robert Penn Warren: A Biography.* "He gave her one, and when she finished the review she went out and sold the book for supper money."

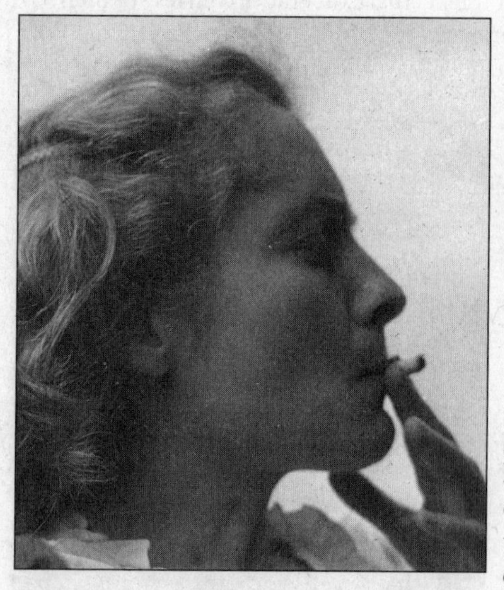

Eleanor Clark smoking on the rooftop of 441 East Ninety-second Street apartment building, New York City, 1940s (Photograph by Walker Evans. Reproduced by permission of the Metropolitan Museum of Art, Anonymous Gift, 1999 [1999.246.79]. Copyright © by the Walker Evans Archive, the Metropolitan Museum of Art.)

Eleanor's early political sympathies lay with the Trotskyites. Accordingly, she traveled to Mexico and offered her services to the exiled Trotsky, who headquartered his movement in a villa on loan to him from the parents of Frida Kahlo. There, Eleanor translated French documents for the anti-Stalinist leader and married his Czech secretary, Jan Frankel. It was a short-lived marriage, later described by Eleanor as a union intended to produce one thing only: a visa.

Though perhaps most identified with *The Partisan Review,* Eleanor also worked as an editor at W. W. Norton & Company. During World War II she was employed at the "Office of Strategic Services monitoring French and Italian publications and interviewing refugees" (Blotner). Always she pursued her fiction and nonfiction—a pursuit which over the years found her writing from places of intense rustication: a barn at the Pennsylvania home of novelist Josephine Herbst (1936); "a donkey's winter stable" in Ile de Port-Cros, France (1965); a barn in Grenoble, France (early 1970s); an old hunter's cabin in Vermont (1970s and 1980s).

Eleanor's sharp mind and good looks were unignorable. Saul Bellow, speaking at her 1996 memorial service, recalled "a breezy young woman with a fine figure, attractive, a lively conversationalist, a great asker of difficult questions." Shirley Hazard and John Kenneth Galbraith and registered similar descriptions at the memorial service. Said Hazard: "Her very appearance showed disparity: the wilderness woman in dungarees could transfigure herself into a distinguished beauty decorously dressed—as one sees in a heartshaking photograph on the flap of *Rome and a Villa*. . . . Eleanor is wearing a low-cut dress and a fine Greek necklace. Looking into those eyes one feels that the finery has been placed on a lioness, or over a flame." Galbraith: "As a critic, Eleanor was not disposed to be tolerant. If something was pretentious, mediocre, or bad, she said so and with emphasis. On several occasions . . . I ventured approval where she did not think it was merited. She left me in no doubt as to my error. She was a valued and faithful member of the American Academy of Rats and Letters, and I remember, as will others, the force of her approval of proposed members and even more vividly her voice in rejection."

Robert Penn Warren, La Rocca, Italy, circa 1955
(Photograph courtesy of Rosanna Warren and Gabriel Warren)

Indeed, stories abound—generally conveyed with something approaching good humor—about one of Eleanor's most noticeable habits: her rapid, almost warlike correction of those whose politics, penmanship, or pronunciation of foreign languages conflicted with her own general notions about these things. ▸

Irish poet Louis MacNeice, practically witch-charmed by these forceful qualities, fell in love with Eleanor during his 1939 lecture tour of the United States. No sooner had he boarded ship to return home than he sat about writing a twenty-four-page letter to her: "The chief things are (a) that I love you & (b) that you have got to love me & (c) that if (a) & (b) are established, [(a) is anyway] things will be lovely & there is nothing really—except in the short run to be sad about" (Jon Stallworthy, *Louis MacNeice: A Biography*). In a later letter he wrote: "I have *never* felt like this about anyone, it pervades everything. If I had imagined ideal people for years, I couldn't have come near it." MacNeice would again court Eleanor during a later visit, and would go so far as to dedicate a book to her. She did not, in the end, share his enthusiasm for their relationship.

Eleanor first met her second husband in 1944. Robert Penn Warren, affectionately known as "Red," would go on to become the only person to win Pulitzers for both fiction and poetry (he won two Pulitzers in the latter category), and to become the nation's first poet laureate ("If America is to choose a poet," said Seamus Heaney, "I don't think it could have chosen better"). The first meeting between the redheaded Kentuckian and the blonde New Englander reputedly took place at one of Katherine Anne Porter's parties. Red, "in the last stages of an unhappy marriage" (*New York Times*), went on to see Eleanor only fleetingly over the next seven years. In 1951, however, the now divorced Red invited Eleanor to dinner in New York. "That dinner was followed by others," wrote Joseph Blotner. "When he took her home one early April night she asked him in, and they walked up the four steep wooden flights of stairs. He sat in the comfortable chair, she on the stool before the little Franklin

Eleanor and Rosanna, La Rocca, Italy, circa 1955
(Photograph courtesy of Rosanna Warren and Gabriel Warren)

stove that spread its warmth in the cold room. They chatted and then fell silent. She was staring into the fire. Then he said, 'I wonder how it would be if I kissed you.' She said, 'Well, you could try and see.' "

Red soon moved into Eleanor's tenement apartment on East Seventy-fifth Street, where she lived for about seven years at a rate of twenty-three dollars per month. Red was forty-seven, Eleanor thirty-nine—and pregnant.

The couple bought an old farm in Fairfield, Connecticut, and occupied a rented cottage while Red worked to prepare their new home. One night Eleanor went into labor and gave Red the almost unequalled thrill of delivering his first child, Rosanna, on the dining room floor.

Newsweek described the legendary dwelling almost thirty years later: "Their colonial-era house, hidden from the road by a screen of trees, is actually two converted, adjoining barns, full of books and homey in feeling." It was here in Fairfield that Red and Eleanor raised their children (Gabriel arrived two years after Rosanna), and, of course, pursued their separate writing careers—each of them working from 9 A.M. to 2 P.M. Here, too, they hosted their celebrated black-tie parties each Christmas—parties attended religiously by John Cheever, Walker Evans, Ralph Ellison, and many other artists. Accounts of these parties are especially glowing:

Eleanor dancing, 1960 (Photograph by Walker Evans. Reproduced by permission of the Metropolitan Museum of Art, the Walker Evans Archive, 1994 [1994.253.738.1–284]. Copyright © by the Walker Evans Archive, the Metropolitan Museum of Art.)

- "Eleanor was so gorgeous . . . just a stunning face, and she had a habit of wearing a red velvet dress so you'd know it was Christmas. I will never forget her coming down the stairs in that gorgeous dress" (Alex Szogyi, in *Robert Penn Warren: A Biography*).
- " [P]eople drank so much at the Warrens' party that nobody would have noticed if your fly was undone" (Benjamin Cheever, in the *New York Times*). ▸

Meet Eleanor Clark ***(continued)***

- "During the year when the Warrens hosted their famous black-tie parties, Fanny and I were often among their weekend guests. Marked by good food, fine drinks, and live music for dancing, these were pleasurable occasions of a truly rare order. We were introduced to an array of people—writers, artists, curators, publishers, academics—whom otherwise we might not have encountered, [and] as far as we were aware no other writers gave parties that encompassed such a diversity of backgrounds and talent." (Ralph Ellison, in *Robert Penn Warren: A Biography*).

Rosanna Warren retains vivid memories of the annual event. "Walker and Isabelle Evans always used to stay in my bedroom," she says. "I had a large dollhouse by the bed, and I was annoyed that I would find cigarette butts in the dolls' swimming pool the next day when I repossessed my room.

"My brother Gabriel and I remember those parties from the child's point of view—the house suddenly aswirl in fantastically dressed up grown-ups, people coming from Washington, DC, and New York to spend the night—the long lazy breakfast the next day—the house abundant in bowls of fruit and nuts, my mother and father spending days before tacking up wreaths of evergreens on the beams of our house."

In between these Christmas parties Eleanor and Red, both ruggedly devoted to outdoor pursuits, honed their pastoral existence. "My mother wore blue jeans and baggy men's shirts," says Rosanna. "One day Robert Redford came to discuss a possible film version of my father's novel *A Place to Come To.* Our cleaning lady threatened to quit if my mother didn't go upstairs, brush her hair, and put on some different clothes for Mr. Redford.

"In the afternoon, as like as not my mother would be working in the garden, tamping mulch around plants, moving stones, and getting her fingers very dirty. She was a straight-backed proud Yankee, whose virtues of thrift and honesty (pronounced in a Depression child who as a young woman had lived for years on her own on threadbare means) complemented my father's Southern old-fashioned country courtesy and stoicism. They were a well-tuned couple, both country people by birth, proud of hard, honest work, and contemptuous of what they saw as modern self-indulgence and weakness. The hard work could be on the land or in the study. What mattered is that it was honorable and unpretentious."

Red, who taught at Yale, encouraged students to visit the farm. "My parents made a big point in their lives of befriending younger writers, artist, and scholars," says Rosanna. "The house often had such visitors—not at fancy parties but simply folded into family life."

While Eleanor wrote fiction, it is her nonfiction for which she is justly remembered. Her nonfiction books are fiercely original; each finds the author poised between a rare faith in emotive response and an unflinching commitment to critical thought.

Saul Bellow recalled the occasion of Eleanor's greatest success as a writer: "The three of us went uptown together in a car when Eleanor won a National Book Award for *The Oysters of Locmariaquer.* I took the fiction prize in that year, and Red said to the two of us, 'Enough of this. You've got your medals. Now get out while the getting's good' " (*A Memorial Service for Eleanor Clark*).

Eleanor's nonfiction made quite a ripple:

- "Really how the woman writes, a little shimmer of light comes up from the page" (Katherine Anne Porter).
- "[P]erhaps the finest book ever to be written about a city" (Anatole Broyard in a *New York Times* review of the 1975 reissue of *Rome and a Villa*, first published in 1952).
- "She . . . has an almost Flaubertian conscience about it" (Louis MacNeice).
- "[P]ossibly the most remarkable document of all is *Eyes, Etc.: A Memoir.* Clark . . . was faced with macular degeneration, a drastic impairment of her vision, and wrote this book with a Magic Marker on a large drawing pad, so she could see what she'd set down. It's a strong book about many facets of life, neither self-pitying nor bitter" (Leonore Fleischer, in a *Washington Post* article about the best memoirs of 1977).

As Eleanor's eyesight diminished—owing to the abovementioned disease—she remained "a granite-solid, no-nonsense Yankee" *(Newsweek).* Red settled into a ritual of reading to her from ancient texts, among them Homer and the Bible. Thus did they continue for many years.

Red died in 1989, aged eighty-four; Eleanor in 1996, aged eighty-two.

A Conversation with the Author's Daughter, Rosanna Warren

Rosanna Warren, winner of many poetry awards, including the Lamont Poetry Prize, is the Emma Ann MacLachlan Metcalf Professor of the Humanities at Boston University.

How did your mother end up in Locmariaquer?

She had been in Brittany as a child. Her mother took her two little girls to Brittany and to other parts of France and Italy on money that was cadged from grandparents. In summers she would place them with families. So my mother and her sister, when they were little girls, were placed in a family in Brittany, and their mother would go off, I don't know, studying French or something. And in winters she'd place them in some really horrendous, cold nunneries where my mother nearly died of whooping cough, but the girls survived and they learned excellent French.

When, and how, did she decide to write the book?

She'd been haunted by the landscape of Brittany. In the summer of 1961, when I was seven and my brother was five, she and my father found a house to rent in the country there, in Locmariaquer, and we went off to Brittany. She wasn't planning to write a book about Brittany—it was just that they found a house there.

Did she write about the oysters in order to write about the people? Or was it quite the reverse?

Well, I think that she found the whole world, an interlocking world of culture and mythology and the landscape and the way the people made their living from the oysters, and she became obsessed with it. She was working on her novel *Baldur's Gate* at the time, and she set the novel aside to write *The Oysters.*

Was she blocked when she was writing the novel, or was she simply that taken with the oysters?

She became so excited and enthralled with the Breton landscape and with the role the oysters played in it. I remember, because my brother and I went with her a lot, that she visited oyster farms. In a way she sort of lived at the oyster farms. She was there all the time, talking to the oystermen and tramping around the countryside, and my brother and I would trail along the beach, skip oyster shells on the water, and look at boats while she was having her long conversations. That summer was all oysters, oysters, oysters.

Was your mother openly interviewing people for her book, or was she conversing with them and later jotting down her notes?

Well, I was just seven at the time so I wasn't observing very closely what she was doing. But I think what happened was that she slowly grew into the project. As far as I know it was unplanned. It sort of grew, and the more it grew, the more focused she became about visiting these oyster people. As she became more aware of what she was doing, she wasn't hiding it from them. Maybe she first thought she would write an article, but the work accelerated; the density and richness of that ▸

A Conversation with the Author's Daughter, Rosanna Warren ***(continued)***

culture, that deeply mythological Breton culture, clearly caught her imagination.

Did she write most of the book on site? How long were you there?

We were there for about five months. I remember my parents had taken us out of school early and put us into a little Breton village school.

Really. What was that like?

My brother and I had the rather frightening experience of being in the village school with farm children who mostly spoke Breton. The school was run by nuns, and the nuns were quite violent; I remember them smacking the little farm children if they spoke Breton because they were meant to speak French. I mean, I remember just cowering under my desk because I didn't speak Breton *or* French.

What were your mother's writing habits? Did she rise early to write?

She was very, very regular. She would get up around seven, have breakfast, take care of us children, get us off to whatever we were doing. By nine she was in her study. We actually had two different houses that summer. First we were in a farmhouse that was a little inland, and then, for whatever reason, my parents rented the ground floor of a rather noble house that was perched on a high hill that sloped down to a muddy inlet of the sea. There weren't oyster flats right below that house, but they were very nearby on the coast. I'd say she moved closer to the oysters. In each of these houses she would have a room they would just take over and make into a study. She worked from nine to two nonstop. Then she and my father would have lunch at two, and after two they would ramble around or she would go exploring and talk to people.

It occurred to me that John McPhee's book* Oranges, *published eleven years after* Oysters, *shares similar aims. For instance,* Oranges *describes something many of us take for granted, observes the whole sweep of its subject's history, and depicts the lives of those upon whose labor we depend (in this case the orange growers and pickers). It's a wonderful approach, and a not uncommon one these days, but it must have seemed fairly odd when* Oysters *was published. Do you recall whether this was the case?

I do recall that. I remember a lot of conversation in the house about it, and conversation with her publishers— conversation that we children would overhear when the grown-ups were talking. Publishers didn't even know what to call the book. They didn't exactly have a category for it and bookstores were saying, Well, where we do put it on the shelves? It's not history, it's not travel literature, it's not religion, and it's not a novel. So I think that even when it got the National Book Award there continued to be conversation, people continued to ask, What is this book? What's its genre? ▸

A Conversation with the Author's Daughter, Rosanna Warren ***(continued)***

Was your mother, then, surprised by the book's success?

I think so. Yes, very.

And your father . . . what was his opinion of the book?

My father was immensely proud of my mother and happy for her. He didn't have a particular connection to France before marrying her, and was delighted to be dragged off on these adventures. He would just work on his own things and let her do what she was doing. In the evenings, the family evenings around dinner, there would be this sort of excited chatter about what had happened during the day and who had seen what, and the oysters were part of that larger, exuberant sense of being in a new world.

Did your parents edit one another's work?

No, absolutely not. To survive as writers there had to be a firewall. In fact, they wouldn't even read each other's work before it was published. It was a very stern, strict rule. It isn't easy for two strong writers to live intimately together. And so they made it work by keeping those boundaries.

I found two signed first editions of* The Oysters of Locmariaquer *online—one of them selling for one hundred seventy-five dollars, the other for sixty-five dollars. The latter was inscribed "For Barbara / with good wishes / Eleanor / April, 1964." According to

the seller, the bookplate was that of the poet Barbara Howes. I'm curious, was Barbara Howes a friend?

That's interesting. Barbara was a good friend. She was the ex-wife of the poet William Jay Smith. She lived in Pownal, Vermont, and was sort of a neighbor of ours. She was about an hour and a half away. My parents and Barbara often visited back and forth in Vermont. There was a deep old connection there.

Your mother brought to serious subjects a lively, at times unbridled, use of language—one that calls to mind the James Agee of **Let Us Now Praise Famous Men.** ***For instance, she describes a ninety-six-year-old woman who sits "deposited under a tree for the morning, like a huge damp mattress forced into a garment of black cotton." The way she finds and describes these people in their humble dwellings—again, it really reminds me of what James Agee was doing: there's a daring to her prose. Was that something that distinguished her writing in general? Was her spoken language as provocative as her prose?***

Her spoken language was very salty, lively, and acute. She really was a stylist. In the words that you just read—those are an artist's words and I'm delighted to hear you read them out loud.

Even when I'm enjoying the prose sentence by sentence, paragraph by paragraph, and your mother has elevated her words to an almost fever pitch, there comes a sentence ▸

A Conversation with the Author's Daughter, Rosanna Warren ***(continued)***

that jumps out and catches me unawares. That's one of the things I enjoy about Eleanor's style: these gradations of intensity.

You said that very well. I also think that she learned to establish a special flow of energy between her inner world and her equally powerful sense of an outer world. So on the one hand, her prose in the nonfiction books *Rome and a Villa* and *The Oysters of Locmariaquer* is very subjective prose: you're being shown the world through a powerful temperament. On the other hand, that temperament is being challenged by things that are external to itself, strange to it—which calls forth an even greater intensity of observation and language. This seems to me part of the force of those two books: that there is a compelling interiority to the vision, but it is in response to something that is acknowledged to be outside the self.

Did your mother maintain a correspondence with anyone from Locmariaquer after you'd returned Stateside?

She did maintain correspondence with some of the key people there for a while.

Did she return to Locmariaquer?

We didn't go back and live there. We went to other parts of France. Once the book was written—not that she ever got it out of her system—she was on to other things. She had accumulated a sort of library of books while we were there—I love those books and have some of them. They are all in French and covered with her notes. There are books about Breton mythology, about oysters, about the Breton landscape, the Breton folklore, and the Breton language. It's very moving to me to have inherited these books.

Have you yourself returned to Locmariaquer as an adult? How did you find it?

Yes, I have returned there. I went back there with my children and my husband about twenty years ago. Interestingly enough, I'm writing a book about a Breton poet, Max Jacob. We went to spend a summer on the coast not too far from Locmariaquer, because that was where this poet had grown up. And so in my young womanhood I found myself going back to my childhood landscape, my mother's oyster landscape, and I took my children and my husband to Locmariaquer and we looked at the menhirs and dolmens and ate oysters. It was quite an extraordinary visit for me, with all those layers of memories.

And had the area changed much?

That whole coast of Brittany is more developed now. The French themselves have developed it as more of a tourist area for French summer tourists, but it hasn't destroyed the landscape.

When you were there as a girl what effect did the ancient monuments—the menhirs and dolmens—have on you?

Oh, they're completely haunting. They had a life-changing, life-directing effect on my ▶

A Conversation with the Author's Daughter, Rosanna Warren ***(continued)***

brother. He became a sculptor, and in a large sense all of his adult work is in some way related to that childhood vision of the menhirs and dolmens. As an adult he went and looked at the Neolithic stone circles. He tracked them down in the Orkney Islands and, of course, at Stonehenge in England, and a great part of his art has come out of thinking about those relations between Neolithic monuments and the stars.

That's fascinating. I mean, you're writing a biography of a Breton poet and your brother is sculpting works inspired by his experience of ancient monuments on the coast of Brittany—that area had quite an effect on you.

On the whole family. My father would hole up reading French novelists. I remember him reading Zola and Balzac. And my brother was taking in the dolmens and the menhirs. And I was, without thinking about it, absorbing that Breton landscape and those myths and that language.

In a 1990 **Washington Post** ***article you mention having reviewed your mother's book for your family newspaper. Is this true?***

[Laughs] Yes, if you can call it a review, since I hadn't even read the book—well, I'd read parts of it, but I thought I knew it by intuition.

Was it a friendly review?

Oh yes. It was most enthusiastic.

What was the newspaper called?

It was called *The Family Racket.* It was a newspaper I had founded and of which I was pretty much the sole author. I took it very seriously and went around interviewing everybody. There were articles on what our cats were doing and on what was happening to our pine tree in the front yard. Occasionally the newspaper would take note when one of the grown-ups in the house had produced a book.

Your mother researched her subject extensively. Was it at all insufferable to eat oysters with her after her book's publication?

No. In the first place, we didn't eat oysters all the time. We were living in the United States. She had become snobby about oysters and only wanted to eat— It isn't true that she only wanted to eat Breton oysters, but she became knowledgeable about North American oysters and would distinguish between them. I was just a kid and didn't particularly like eating oysters. I just regarded them as one of my mother's many grown-up passions that I was happy to let her have.

Were there any family jokes about oysters?

There must have been. I just remember general good humor about it. It was delightful to see what had been a summer adventure grow up into a book, and then the book win such an audience.

A Poem by Robert Penn Warren

Praise

I want to praise one I love,
Instructress in the heart's glory,
Who whirls through life in a benign fury,
Or walks alone in the high pine grove.

I want to praise one who sheds light
In darkness where the foot can find
No certainty, and the unlit mind
Wavers, and cannot stand upright.

I want to praise one whose angry joy
Is innocence by wisdom hued,
And whose laughter, at its gayest, is brewed
By knowledge the world is only a toy

To all who can summon courage to live
In innocence and pitying rage
At a sickly and self-pitying age
That scorns the true good the world can give.

Eleanor seated with beverage, Italy, circa 1955 (Photograph courtesy of Rosanna Warren and Gabriel Warren)